WEF/ABC

Wastewater Laboratory Analysts'

Guide to Preparing for the Certification Examination

2000 First printing
2002 Second printing

Water Environment Federation
601 Wythe Street
Alexandria, VA 22314-1994
www.wef.org

Copyright © 2000 by the Water Environment Federation.
Permission to copy must be obtained from WEF.
All Rights Reserved.

ISBN 1-57278-164-5

Table of Contents

	Page
Association of Boards of Certification Education and Experience Requirements	1
Class I	1
Class II	1
Class III	1
Class IV	2
Additional Information Regarding Education and Experience	2
Taking a Certification Examination	4
Be Prepared	4
Taking the Examination	5
Calculation Problems	6
Helpful Hints	6
Examination Checklist	7
Formulas	8
Conversion Factors	11
Abbreviations and Acronyms	12
Practice Questions and Answers	17
Acidity/Alkalinity/pH	17
Questions	17
Answers	30

Biochemical Oxygen Demand/Carbonaceous Biochemical Oxygen Demand/Chemical Oxygen Demand/Dissolved Oxygen   **42**

    Questions   42
    Answers   53

Chlorine/Ozone   **63**

    Questions   63
    Answers   66

Microbiological Examination   **69**

    Questions   69
    Answers   80

Process Control   **90**

    Questions   90
    Answers   95

Residue   **99**

    Questions   99
    Answers   107

General   **113**

    Questions   113
    Answers   122

Safety   **130**

    Questions   130
    Answers   136

Laboratory Apparatus/Reagents/Techniques   **141**

    Questions   141
    Answers   151

Mathematics   **159**

    Questions   159
    Answers   172

Sampling   **190**

    Questions   190
    Answers   197

Compliance	**202**
Questions	202
Answers	204
Conductivity	**205**
Questions	205
Answers	208
Quality Assurance/Quality Control	**210**
Questions	210
Answers	221
Nitrogen	**230**
Questions	230
Answers	233
Oil/Grease	**236**
Questions	236
Answers	238
Phosphorus	**240**
Questions	240
Answers	241
Inorganics	**242**
Questions	242
Answers	245
Organics	**248**
Questions	248
Answers	250
Glossary	**253**

References 279

Appendix A Table of Elements 281

Appendix B   Approved Methods 283

Water Environment Federation

Founded in 1928, the Water Environment Federation is a not-for-profit technical and educational organization. Its mission is to preserve and enhance the global water environment. Federation members number more than 38 000 water quality professionals and specialists from around the world, including engineers, scientists, government officials, utility and industrial managers and operators, academics, educators and students, equipment manufacturers and distributors, and other environmental specialists.

>For information on membership, publications, and conferences,
>contact

Water Environment Federation
601 Wythe Street
Alexandria, VA 22314-1994 USA
(703) 684-2400
http://www.wef.org

Association of Boards of Certification

Established in 1972, ABC is a not-for-profit organization through which certification boards may communicate and, as a result, better discharge their responsibilities. The Association of Boards of Certification assists certification authorities in developing stronger administrative programs and more effective criteria for assessing an analyst's qualifications. It seeks to improve analyst certification laws, their administration, and their effectiveness; promote analyst certification as a means of ensuring effective data; define and maintain nationally recognized qualifications for analyst certification; promote uniformity of qualifications, standards, and practices in analyst certification; facilitate interboard transfer of certification for qualified individuals; and assist newly created boards in establishing initial policies and procedures.

Preface

The Association of Boards of Certification (ABC) and the Water Environment Federation (WEF) worked together to produce this document based on validated "need-to-know" job analyses from the ABC Testing Service. The document is designed to give the reader practice answering multiple-choice examination questions. The questions test the various levels of skill and knowledge required by a laboratory analyst working at a wastewater treatment plant. The guide is also intended as a reference tool for laboratory analysts during examination preparation and on a day-to-day basis. In addition to listing the ABC education and experience requirements and information regarding taking a certification examination, this guide also includes all of the formulas used in the wastewater treatment plant laboratory, including conversion factors and a detailed list of abbreviations and acronyms.

The questions included in this guide have been chosen to sample as many different aspects of an analyst's job as possible. However, because of the tremendous variety in equipment, processes, conditions, and analyst duties, not all of the questions may be useful in all possible certification applications. The questions and answers themselves are organized by topic. Detailed solution sets are provided for those answers requiring mathematical computation. A reference is provided for each question and answer, with a comprehensive reference list at the end of the guide. The table of elements and tables of approved methods make up the appendices.

The following people participated in the development of *WEF/ABC Wastewater Laboratory Analysts' Guide to Preparing for the Certification Examination*.

Chair

Kathleen M. Cook

Authors

Leonard W. Ashack, III	Janice O' Byrne
Lowell Malcolm Baker	James G. Poff
Laura Conrad	Roger Rardain
Charlene Givens	Roy-Keith Smith
Martie Groome	Mark Stoffan
Sara Lewis	Marguerite Tanner
William M. McElhaney	

Reviewers

Sandra Arnold	Richard G. Weigand
Lowell Malcolm Baker	Dave Wetzel
Tracy Baxter	Jane Winkler
Frank Barosky	Mark Wyzalek
Debra Clark	
Laura Conrad	
Margaret Doss	
Ernest U. Earn	
Hsiao Yung Ford	
Mike Heniken	
John Hoffman	
Nanette Holdek	
Sylvia McCrary	
Doug Mercer	
Mary Anne Rose	
Pam Schweitzer	
Keniz Siddiqui	
Roy-Keith Smith	
Mark Stoffan	

ASSOCIATION OF BOARDS OF CERTIFICATION EDUCATION AND EXPERIENCE REQUIREMENTS

The education and experience requirements for wastewater laboratory analysts are as follows:

CLASS I

- High school diploma, general equivalency diploma (GED), or equivalent and
- 1 year of acceptable wastewater laboratory experience.

Note that no substitution for education or experience shall be permitted.

CLASS II

- High school diploma, GED, or equivalent and
- 3 years of acceptable wastewater laboratory experience.

Note that a maximum of 675 contact hours or 68 continuing education units (CEUs) or 68 quarter credits or 45 semester credits of post-high school education in the environmental control field, engineering, or related science may be substituted for 1 1/2 years of wastewater laboratory experience.

CLASS III

- High school diploma, GED, or equivalent;
- 4 years of acceptable wastewater laboratory experience; and
- 900 contact hours or 90 CEUs or 90 quarter credits or 60 semester credits of post-high school education in the environmental control field, engineering, or related science.

Note that a maximum of 900 contact hours or 90 CEUs or 90 quarter credits or 60 semester credits of post-high school education in the environmental control field, engineering, or related science may be substituted for 2 years of wastewater laboratory experience. A maximum of 1 year of experience in a Class II or higher position may be substituted for 450 contact hours or 45 CEUs or 45 quarter credits or 30 semester credits of post-high school education in the environmental control field, engineering, or related science.

CLASS IV

- High school diploma, GED, or equivalent;
- 4 years of acceptable wastewater laboratory experience; and
- 1800 contact hours or 180 CEUs or 180 quarter credits or 120 semester credits of post-high school education in the environmental control field, engineering, or related science.

Note that a maximum of 900 contact hours or 90 CEUs or 90 quarter credits or 60 semester credits of post-high school education in the environmental control field, engineering, or related science may be substituted for 2 years of environmental laboratory experience. A maximum of 2 years of experience in a Class III or higher position may be substituted for 900 contact hours or 90 CEUs or 90 quarter credits or 60 semester credits of post-high school education in the environmental control field, engineering, or related science.

ADDITIONAL INFORMATION REGARDING EDUCATION AND EXPERIENCE

- Education applied to the experience requirements shall not also be applied to the education requirements.
- Experience applied to the education requirements shall not also be applied to the experience requirements.
- Where applicable, related experience in operations, maintenance, other environmental control utility positions, and allied trades such as a hospital laboratory technician or other certification categories may be substituted for one-half of the experience requirement.

- The maximum substitution of education and related experience for wastewater laboratory experience shall not exceed 50% of the stated wastewater laboratory experience requirement.

TAKING A CERTIFICATION EXAMINATION

Taking and passing a certification examination is an important part of a wastewater laboratory analyst's job. Here are some helpful hints for taking a certification examination.

BE PREPARED

The first step in passing a certification examination is to prepare for the test in advance by studying the type of information that will be included on the examination. One method of preparation for the test is to attend training classes offered by local wastewater utilities, community colleges, and vocational–technical schools. Another useful method is to read and study reference books on wastewater treatment. In addition, some state or provincial environmental departments provide sample tests, lists of material covered in the examinations, or examples of typical test problems.

Certification examinations are intended to test every aspect of a laboratory analyst's involvement in wastewater treatment. Because calculations are often involved, certification examinations include mathematical problems dealing with detention time, flow rates, chemical dosages, tank volumes, and so forth. To prepare for mathematical problems, it is a good idea to memorize the more common conversion factors used in wastewater treatment mathematics.

Remember that examinations are changed frequently. Thus, do not try to learn the answers to specific problems that you may have heard about from other analysts. Concentrate on learning how to solve various types of problems, such as how to prepare a biochemical oxygen demand test or how to conduct a bacteriological examination. In this way, you will be prepared to handle any question on the test, not just certain questions you have memorized.

Although laboratory analysts are responsible for knowing all aspects of wastewater laboratory work, certification tests in different regions emphasize some aspects more than others. This guide is broken into the topics that are typically emphasized on the examinations. By talking with other analysts who have already taken the test, you can learn which additional areas may be emphasized on the examination given in your area.

Finally, allow yourself enough time to prepare for the examination. Do not try to study all of the information a few days before the test. Cramming is not typically beneficial and can lead to confusion and exhaustion.

TAKING THE EXAMINATION

On the day of the examination, be at the testing location well ahead of the announced starting time. Be sure that you have the necessary materials, such as a spare pencil and your admittance slip (if required). Also, remember to put fresh batteries in your calculator because working mathematical calculations by hand can use valuable time. Once you are given the test, skim through the questions to get a general idea of the kinds of questions that you will be answering. By seeing the types of questions and mathematical problems you will be answering, you will have a better idea about how to use your time. It is a good idea to answer all of the easy questions first and then solve the difficult problems that will take more time to complete.

If the test has essay questions and lengthy mathematical problems, think about the person who will grade these questions. To help the grader, write your answer clearly and in an order that the grader can easily follow. Remember that graders will be reviewing lots of essays and calculations. Although they may not intend to lower your grade for sloppy or unreadable work, such work certainly will not help your score. Underline answers to mathematical problems—do not make the grader search through pages of figures to find your answer.

CALCULATION PROBLEMS

Because calculation problems are an important and often difficult part of certification examinations, several guidelines should be followed when solving calculation problems.

(1) Read each question carefully to be sure that you know what answer is required.
(2) Make a drawing or sketch if it will help to solve the problem.
(3) Simplify the problem. If the problem is complex, break it down into small pieces you can solve separately.
(4) Make all necessary conversions first (such as converting 1 minute to 60 seconds, if the question requires an answer in seconds).
(5) Be sure the decimal point is in the right place.
(6) Check to see if your answer makes sense. For example, if you get an answer of 200 000 mg/L instead of, say, 2000 mg/L for a question that asks for the suspended solids concentration in a biological reactor, your answer is obviously wrong.
(7) Be sure that your answer is expressed in the correct units.
(8) After completing the calculation, rework the problem again to double-check your answer.

HELPFUL HINTS

Most certification examinations are primarily composed of multiple-choice questions. Even if you are unsure of the answer, you should answer all questions. Often, you can determine that one or two of the potential answers are obviously wrong. If you narrow your choice to only two possible answers, you have a 50–50 chance of getting the right answer, no matter which answer you choose.

Take the examination methodically and deliberately. Avoid the natural temptation to rush through it. Be relaxed while taking the examination, and take a break periodically to ease the tension.

EXAMINATION CHECKLIST

Before leaving your home or workplace to go to the testing location, be sure that you have the following items so that you will be well prepared to take the examination:

- An admittance slip (if one is required),
- Two sharpened No. 2 pencils with good erasers,
- A calculator,
- Fresh batteries for the calculator,
- A watch,
- Glasses (if you wear them), and
- Confidence.

Remember, if you have properly prepared for the examination and can perform your work well, you should pass the examination.

FORMULAS

Acidity, mg/L = $\dfrac{\text{Titrant Used, mL} \times \text{Normality of NaOH Titrant} \times 50\,000}{\text{Sample Volume, mL}}$

Alkalinity, mg/L = $\dfrac{(\text{Volume of Titrant, mL})(\text{Normality of Titrant})(50\,000)}{\text{Sample Volume, mL}}$

Area of Circle = $\dfrac{(\pi)(\text{Diameter}^2)}{4}$ or $(\pi)(\text{Radius}^2)$

Area of Cylinder = $(\pi)(\text{Diameter})(\text{Height})$ or $(2)(\pi)(\text{Radius})(\text{Height})$

Area of Rectangle = (Length)(Width)

Area of Triangle = $\dfrac{(\text{Base})(\text{Height})}{2}$

BOD = $\dfrac{(\text{Initial DO, mg/L} - \text{Final DO, mg/L})}{\dfrac{\text{Sample Volume, mL}}{\text{Bottle Volume, mL}}}$

Calculated Recovery, mg/L = Concentration of Sample with Spike, mg/L − Concentration of Sample, mg/L

COD (Titrimetric Method) =

$\dfrac{(\text{FAS Used for Blank, mL} - \text{FAS Used for Sample, mL}) \times \text{Molarity of FAS} \times 8000}{\text{Sample Volume, mL}}$

Circumference of Circle = $(\pi)(\text{Diameter})$

Coliform Density per 100 mL = $\dfrac{\text{Coliform Colonies Counted} \times 100}{\text{Filtered Sample, mL}}$

$$\text{Composite Sample Single Portion} = \frac{(\text{Instantaneous Flow})(\text{Total Sample Volume})}{(\text{Number of Portions})(\text{Average Flow})}$$

Dilution = (Initial Volume) (Initial Molarity) = (Final Volume) (Final Molarity)

$$\text{Efficiency (Percent Removed)} = \frac{(\text{Quantity In} - \text{Quantity Out})}{\text{Quantity In}} \, 100\%$$

$$\text{F:M} = \frac{\text{Amount of BOD}}{\text{Amount of MLVSS}}$$

Geometric Mean = $1 \times 10^{(\text{sum of log 10 of results/number of results})}$

$$\text{Hardness} = \frac{(\text{Volume of Titrant, mL})(B)(1000)}{\text{Volume of Sample, mL}}$$

Where

B = Amount of $CaCO_3$ (mg) = equivalent to 1.00 mL EDTA titrant

$$\text{MCRT} = \frac{\text{SS in the Process, kg (lb)}}{\text{Amount of SS Wasted, kg (lb)/d} + \text{Amount of SS in Effluent, kg (lb)/d}}$$

Milliequivalent = (1/1000 Equivalent)

$$\text{Molarity} = \frac{(\text{Weight of Compound, g})/(\text{Molecular Weight of Compound, g/mol})}{(\text{Volume of Solution, L})}$$

$$\text{Molarity} = \frac{\text{Normality}}{\text{Number of Replaceable Electrons}}$$

$$\text{Normality} = \frac{\text{Weight of compound} \times \dfrac{\text{MW of Compound, g/mol}}{\text{Number of Replaceable Electrons}}}{\text{Volume of Solution, L}}$$

$$\text{Oxygen Uptake} = \frac{\text{Oxygen Usage, mg/L DO}}{\text{Time, h}} = \frac{\text{Initial DO, mg/L} - \text{Final DO, mg/L}}{\text{Time, h}}$$

Percent Accuracy = 100% − [(Known Addition or Matrix Spike − Calculated Recovery) × 100%]

$$\text{Percent Removal} = \frac{\text{Total Weight In, mg} - \text{Total Weight Out, mg}}{\text{Total Weight In, mg}} \times 100\%$$

or

$$\text{Percent Removal} = \frac{\text{Total Concentration In, mg/L} - \text{Total Concentration Out, mg/L}}{\text{Total Concentration In, mg/L}} \times 100\%$$

$$\text{Percent Total Solids} = \frac{\text{Dry Solids Weight, g}}{\text{Wet solids weight, g}} \times 100\%$$

$$\text{Percent Volatile Solids} = \frac{(\text{Dry Solids Weight, mg} - \text{Ash Solids Weight, mg})}{\text{Dry Solids Weight, mg}} \times 100\%$$

$$\text{Population Equivalent} = \frac{(\text{Flow, m}^3\text{/d})(\text{BOD, mg/L})(0.001\,\frac{\text{kg/m}^3}{\text{mg/L}})}{\text{kg BOD/d}}$$

$$\text{Population Equivalent} = \frac{(\text{Flow, mgd})(\text{BOD, mg/L})(8.34\,\frac{\text{lb/mil. gal}}{\text{mg/L}})}{\text{lb BOD/d}}$$

Seeded BOD =

$$\frac{(\text{Initial Sample DO, mg/L} - \text{Final Sample DO, mg/L}) - (\text{Initial Seed Control DO, mg/L} - \text{Final Seed Control DO, mg/L})\,f}{\frac{\text{Sample Volume, mL}}{\text{Bottle Volume, mL}}}$$

$$f = \frac{\text{Volume of Seed in Sample, mL}}{\text{Volume of Seed in Control, mL}}$$

Sludge Age = Same as equation for MCRT

$$\text{SVI, g/mL} = \frac{(\text{Settleable Sludge Volume, mL/L}) \times (1000\,\text{mL/L})}{\text{Suspended Solids, mg/L}}$$

$$\text{Solids Applied, kg/d} = (\text{Flow, m}^3\text{/d})(\text{Concentration, mg/L})(0.001\,\frac{\text{kg/m}^3}{\text{mg/L}})$$

Solids Applied, lb/d = (Flow, mgd) (Concentration, mg/L) (8.34 $\frac{\text{lb/mil. gal}}{\text{mg/L}}$)

Solids Concentration = $\frac{\text{Weight, mg}}{\text{Volume, L}}$

Standard Deviation = Square Root of the Variance

TSS = $\frac{(\text{Weight of Filter and Dry Residue, mg} - \text{Weight of Filter, mg})}{\text{Sample Volume, mL}}$ × 1000 mL/L

Velocity = $\frac{\text{Flow}}{\text{Area}}$ *or* $\frac{\text{Distance}}{\text{Time}}$

Volume of Cone = $\frac{[(\pi)(\text{Diameter}^2)/4](\text{Height})}{3}$ or $\frac{(\pi)(\text{Radius}^2)(\text{Height})}{3}$

Volume of Cylinder = $\frac{(\pi)(\text{Diameter}^2)(\text{Height})}{4}$ or $(\pi)(\text{Radius}^2)(\text{Height})$

Volume of Rectangle = (Length) (Width) (Height)

CONVERSION FACTORS

1 ac = 43 560 sq ft

1 cu ft = 7.5 gal

1 ft = 0.304 8 m

1 gal = 3.785 L

1 gal of water weighs 8.34 lb

1 gr/gal = 17.12 mg/L

1 hp = 0.745 7 KW

1 lb/mil. gal = 0.1198 mg/L

$$1 \text{ M} = \frac{1 \text{ mol}}{1 \text{ L}}$$

1 mgd = 694 gal/min

1 lb = 0.453 6 kg

1% solution = 10 000 mg/L

°C = (°F − 32) (0.555 6)

°F = [(°C) (1.8)] + 32

ABBREVIATIONS AND ACRONYMS

ABC	= Association of Boards of Certification
ac	= acre
$AgSO_4$	= silver sulfate
$Al_2(SO_4)_3$	= aluminum sulfate
$AlCl_3$	= aluminum chloride
B_2O_3	= boric acid
BOD	= biochemical oxygen demand
$Ca(OH)_2$	= calcium hydroxide
$CaCl_2$	= calcium chloride
$CaCO_3$	= calcium carbonate
CBOD	= carbonaceous biochemical oxygen demand
CCl_4	= carbon tetrachloride
°C	= degrees Celsius
CEU	= continuing education unit
CFR	= Code of Federal Regulations
CFU	= colony-forming unit
$CHCl_3$	= chloroform
CH_4	= methane
CO_2	= carbon dioxide
CO_3	= carbonate
$CoCl_2$	= cobalt chloride

COD	= chemical oxygen demand
CS_2	= carbon disulfide
cu ft	= cubic foot
$CuSO_4$	= copper sulfate
d	= day
DO	= dissolved oxygen
DPD	= *N*, *N*-diethyl-*p*-phenylenediamine
EC	= effective concentration
ECD	= electron capture detector
EDTA	= ethylenediamine tetraacetic acid
F:M	= food-to-microorganism ratio
°F	= degrees Fahrenheit
FAS	= ferrous ammonium sulfate
$FeCl_2$	= ferrous chloride
$FeCl_3$	= ferric chloride
FID	= flame-ionization detector
ft	= foot
g	= gram
gal	= gallon
GED	= general equivalency diploma
GGA	= glucose–glutamic acid
gpd	= gallons per day
gpm	= gallons per minute
h	= hours
H_2O	= water
H_2O_2	= hydrogen peroxide
H_2S	= hydrogen sulfide
H_2SO_4	= sulfuric acid
HCl	= hydrochloric acid
HClO	= hypochlorous acid
$HgSO_4$	= mercuric sulfate

HNO_3	= nitric acid
hp	= horsepower
IC	= inhibiting concentration
in.	= inch
KCl	= potassium chloride
$K_2Cr_2O_7$	= potassium dichromate
KH_2PO_4	= potassium dihydrogen phosphate
kg	= kilogram
KI	= potassium iodide
KOH	= potassium hydroxide
kW	= kilowatt
L	= liter
lb	= pound
LC	= lethal concentration; this is usually defined as the median (50%) lethal concentration (LC_{50})
LOEC	= lowest observed effect concentration
m	= meter
M	= molar
MAD	= mass acceleration detector
MCRT	= mean cell residence time
mg	= milligram
mg/L	= milligrams per liter
$MgCl_2$	= magnesium chloride
mgd	= million gallons per day
$MgSO_4$	= magnesium sulfate
mil. gal	= million gallon
min	= minute
mL	= milliliter
MLSS	= mixed liquor suspended solids
MLVSS	= mixed liquor volatile suspended solids
MnO_4	= permanganate

$MnSO_4$	= manganous sulfate
mol	= mole
MPN	= most probable number
mS/m	= millisiemens per meter
MW	= molecular weight
N	= normal
N_2O	= nitrous oxide
Na_2CO_3	= sodium carbonate
Na_2HPO_4	= disodium hydrogen phosphate
$Na_2S_2O_3$	= sodium thiosulfate
Na_2SO_3	= sodium sulfite
Na_2SO_4	= sodium sulfate
NaCl	= sodium chloride
$NaHCO_3$	= sodium bicarbonate
NaOH	= sodium hydroxide
NFPA	= National Fire Protection Association
NH_3	= ammonia
NH_4Cl	= ammonium chloride
NH_4OH	= ammonium hydroxide
$(NH_4)_2SO_4$	= ammonium sulfate
NIST	= National Institute of Standards and Technology
NOEC	= no observed effect concentration
NPDES	= National Pollutant Discharge Elimination System
NTU	= nephelometric turbidity unit
PCB	= polychlorinated biphenyls
π	= 3.141 59
PID	= photoionization detector
SiO_2	= silicon dioxide
SO_2	= sulfur dioxide
sq ft	= square foot
SS	= suspended solids

S.U.	= standard unit
SVI	= sludge volume index
TC	= to contain
TD	= to deliver
TKN	= total Kjeldahl nitrogen
TSS	= total suspended solids
U.S. EPA	= U.S. Environmental Protection Agency
WEF	= Water Environment Federation

PRACTICE QUESTIONS AND ANSWERS

ACIDITY/ALKALINITY/pH

Questions

1. The pH measurement is temperature dependent. Which temperature measurement should always be reported to the nearest degree Celsius for each pH measurement?

    a. Sample
    b. Ambient
    c. Buffer
    d. Equipment

2. Report pH measurements in standard units to the nearest

    a. 0.001
    b. 0.01
    c. 0.1
    d. 1.0

3. Neutral pH at 25 °C is 7.0. Neutral pH of a sample at 0 °C

    a. Increases
    b. Decreases
    c. Stays the same
    d. Fluctuates

4. The acid or basic condition of a sample is expressed by its

    a. Alkalinity

b. pH
   c. Acidity
   d. Conductivity

5. What is the effect of anaerobic conditions on the pH of wastewater?

   a. Buffers the pH
   b. Raises the pH
   c. Lowers the pH
   d. Has no effect

6. The pH reading is considered stable and drift free when

   a. The pH reading begins to slowly rise or drop in the opposite direction
   b. The pH reading changes fewer than 0.5 standard pH units within 60 seconds
   c. The pH reading changes fewer than 0.1 pH units
   d. There is no further change in the temperature reading

7. The most common cause of pH electrode fouling is coating with

   a. Gritty material
   b. Volatile solids
   c. Oily material and/or precipitated solids
   d. Dissolved solids

8. Fresh pH buffer solutions should be prepared

   a. Daily
   b. Weekly

c. Monthly
d. Per manufacturer's specifications

9. What should an analyst do with the fill hole on refillable nonsealed pH electrodes during pH measurements?

   a. Leave it covered
   b. Immerse it in buffer
   c. Immerse it in sample
   d. Leave it uncovered

10. The hold time for a pH sample is

    a. 12 hours
    b. 24 hours
    c. 48 hours
    d. None, analyze immediately

11. To avoid the gain or loss of dissolved gases contributing to the acidity or alkalinity of a sample

    a. Avoid vigorous shaking or mixing of the sample
    b. Bubble air through the solution
    c. Warm the sample to 35 °C
    d. Add a buffer solution

12. Acidity samples containing free chlorine residual are pretreated by adding one drop of

    a. 0.1 N DPD

b. 0.1 N Na₂SO₄
c. 0.1 M Na₂S₂O₃
d. 0.1 N NaOH

13. The endpoint for acidity may be determined by titrating to a

    a. Potentiometric endpoint of 5.5
    b. Phenolphthalein indicator endpoint of 8.3
    c. Metacresol purple indicator endpoint of 4.2
    d. pH of 7.0

14. Using a 50-mL sample, what range does the U.S. EPA Method 310.1 (Alkalinity: Titrimetric) cover

    a. 0.1 to 100 mg/L as CaCO₃
    b. 10 to 1000 mg/L as CO₂
    c. 10 to 1000 mg/L as CaCO₃
    d. 500 to 1000 mg/L as CaCO₃

15. Suspended solids present in the sample or precipitates formed during titration may cause sluggish electrode response. Offset this slow electrode response by

    a. Allowing the sample to sit for 10 minutes between each titrant addition
    b. Stirring the sample faster
    c. Slowly adding drops of titrant as the endpoint pH is approached
    d. Adding some polymer to aid coagulation

16. The acidity of a sample is measured by the quantity of

    a. A strong acid required to bring the sample to a designated pH
    b. A weak acid required to bring the sample to a designated pH

c. A strong base required to bring the sample to a designated pH
d. A weak base required to bring the sample to a designated pH

17. During the acidity measurement, stir the sample with a magnetic stir bar at a rate that

   a. Causes desired sample surface disturbance
   b. Allows the sample to splash excessively, thereby entraining air
   c. Thoroughly but gently mixes the sample and titrant with no splashing
   d. Allows SS to settle

18. The standard titrant for acidity using *Standard Methods* 2310 B. (Acidity: Titration Method) is

   a. 0.02 N NaOH
   b. 0.2 M NaOH
   c. 0.02 N Ca(OH)$_2$
   d. 0.02 N CaCl$_2$

19. The correct preservation and hold time for samples being stored before acidity or alkalinity measurements is

   a. 7 days at 4 °C
   b. 14 days at 0 °C
   c. 14 days at 4 °C
   d. No hold time; analyze immediately at > 4 °C

20. Results of an acidity measurement are reported as milligrams per liter CaCO$_3$ and should include the

   a. Specific pH endpoint used

b. Type of analysis equipment used

c. Color indicator used

d. Sample temperature at time of analysis

21. The pH endpoint for the U.S. EPA Method 305.1 (Acidity) is

a. 3.7
b. 4.2
c. 8.2
d. 8.7

22. The pH endpoints used for a total acidity measurement according to *Standard Methods* 2310 B. (Acidity: Titration Method) are

a. 3.7 and 8.3
b. 4.2 and 8.2
c. 4.2 and 8.7
d. 8.7 and 9.7

23. Which of the following is the correct way to calculate acidity?

a. Acidity, mg/L = (Titrant Used, mL × Normality of NaOH Titrant × 50 000)/Sample Volume, mL
b. Acidity, mg/L = (Sample Used, mL × Normality of NaOH Titrant × 10 000)/Titrant Used, mL
c. Acidity, mg/L = (Titrant Used, mL × Normality of NaOH Titrant × 10 000)/Sample Volume, mL
d. Acidity, mg/L = (Sample Used, mL × Normality of NaOH Titrant × 50 000)/Titrant Used, mL

24. Alkalinity is reported as

    a. milligrams per kilograms CaCO₃
    b. milligrams per liter CO₃
    c. micrograms per liter CaCO₃
    d. milligrams per liter CaCO₃

25. Which of the following is the correct way to calculate alkalinity?

    a. Alkalinity, mg/L = (Acid Titrant Used, mL × Normality of Acid Titrant × 50 000)/Sample Volume, mL
    b. Alkalinity, mg/L = (Sample Used, mL × Normality of NaOH Titrant × 50 000)/Titrant Used, mL
    c. Alkalinity, mg/L = (Titrant Used, mL × Normality of NaOH Titrant × 50 000)/Sample Volume, mL
    d. Alkalinity, mg/L = (Sample Used, mL × Normality of Acid Titrant × 50 000)/Titrant Used, mL

26. Bromcresol green indicator solution has a pH endpoint of

    a. 3.7
    b. 4.3
    c. 4.5
    d. 8.3

27. Soaps, oily matter, suspended solids, or precipitates may coat pH electrodes and cause sluggish response. To overcome these interferences

    a. Pretreat the sample with trichlorotriflouroethane to remove oily residue
    b. Filter the sample before analysis

c. Clean the electrodes occasionally between titrations

d. Wash the electrode with alkaline-based titrant daily

28. Adjust the sample volume used for alkalinity titration to avoid

   a. Using too much of the sample
   b. Having a total titration time greater than 15 minutes
   c. Having a $CaCO_3$ concentration greater than 10 000 mg/L
   d. Having a titration volume greater than 50 mL

29. When using a pH meter and electrode that do not have automatic temperature compensation, adjust the sample temperature before titration to

   a. 20 °C ± 1 °C
   b. 20 °C ± 2 °C
   c. 23 °C ± 2 °C
   d. 25 °C ± 2 °C

30. Select the appropriately sized vessel/container for alkalinity titrations to

   a. Keep the air space above the solution to a minimum
   b. Leave room for a maximum addition of 100 mL of titrant
   c. Provide the largest possible surface area for complete mixing
   d. Allow sufficient space to mix the sample at a high rate of speed without splashing out of the container

31. The standard titrant for U.S. EPA Method 310.1 (Alkalinity) is

   a. 0.01 M EDTA
   b. 0.02 N H_2SO_4 or HCl for alkalinity concentrations <1000 mg $CaCO_3$/L

c. 1.0 N H_2SO_4 or HCl for alkalinity concentrations >1000 mg $CaCO_3$/L
d. 0.1 N $Na_2S_2O_3$

32. The potentiometric endpoint for total alkalinity is a pH of

    a. 3.7
    b. 4.3
    c. 4.5
    d. 8.3

33. Alkalinity is a measure of the

    a. Capacity to neutralize acids
    b. Capacity to neutralize bases
    c. Change in pH in aerobic samples
    d. Increase in salinity

34. Bromcresol green, methyl red, and metacresol purple indicators are used in what test?

    a. Membrane filter test for fecal coliform
    b. Hardness
    c. Alkalinity
    d. MPN test for fecal coliform

35. Compared to a solution with a pH of 4, an acid solution with a pH of 2 is

    a. Twice as alkaline
    b. Twice as acidic
    c. 10 times as acidic
    d. 100 times as acidic

36. What type of sample should be collected for pH testing?

    a. A composite sample
    b. A continuous sample
    c. A grab sample
    d. A regulated sample

37. In the acidity titration method, phenolphthalein is used in the process of titration as a(n)

    a. Coagulant
    b. Indicator
    c. Neutralizer
    d. Oxidizer

38. If a sample of water is known to be basic, which buffers should be used to standardize the pH meter before pH measurement?

    a. 1 and 5
    b. 4 and 7
    c. 7 and 10
    d. 11 and 13

39. A sample has a pH of 6.5 at 25 °C. This sample can be said to be

    a. Acidic
    b. Alkaline
    c. Neutral
    d. Basic

40. The pHs of six grab samples are 6.0, 7.5, 6.9, 5.2, 8.9, and 8.5. From this, we may conclude that the pH of the composite sample is

   a. 4.2
   b. 7.2
   c. 8.9
   d. pH values cannot be averaged

41. In the titration method for alkalinity, the acid titrant is standardized with which solution?

   a. $CaCO_3$
   b. $NaHCO_3$
   c. $NaOH$
   d. Na_2CO_3

42. The definition of pH is

   a. The negative logarithm of the hydrogen ion activity
   b. The positive logarithm of the hydrogen ion activity
   c. The negative logarithm of the hydroxyl ion activity
   d. The positive logarithm of the hydroxyl ion activity

43. Acidity in wastewater can be caused by

   a. Septic anaerobic conditions in the collection system
   b. Use of lime in industrial discharges
   c. Carbonates in the wastewater
   d. Formation of colloidal particles in the collection system

44. What three principle forms of alkalinity are present in many waters?

   a. Carbonate, phenolphthalein, and hydroxide
   b. Carbonate, hydroxide, and bicarbonate
   c. Carbon dioxide, hydroxide, and bicarbonate
   d. Carbon dioxide, carbonate, and phenol

45. When no phenolphthalein alkalinity is found in a wastewater sample, all of the remaining alkalinity present is

   a. Bicarbonate alkalinity
   b. Carbonate alkalinity
   c. Carbon dioxide
   d. Hydroxide

46. Which of the following has the strongest effect on a pH measurement?

   a. Turbidity
   b. Colloidal suspensions
   c. Oxidants
   d. Temperature

47. How many buffers are recommended for proper calibration of a pH meter?

   a. 1
   b. 2
   c. 3
   d. 4

48. How often should a pH meter be calibrated to ensure accuracy?

   a. Twice a day
   b. Daily
   c. Weekly
   d. Monthly

49. Which parameter is responsible for changing the slope of a pH meter calibration?

   a. Conductivity
   b. Resistivity
   c. Temperature
   d. Salinity

50. The concentration of hydrogen ions is most closely related to

   a. Alkalinity
   b. Buffer capacity
   c. pH
   d. Normality

51. The pH endpoint for a total alkalinity test is

   a. 4.5
   b. 5.5
   c. 7.0
   d. 9.0

Answers

1. a. Sample

   Sample temperature affects the pH result and, therefore, should be reported with it, especially if pH-measuring equipment has temperature compensation. Please note that temperatures of the sample, buffer, and meter/probe should ideally be the same during pH analysis.

   Refer to American Public Health Association; American Water Works Association; and Water Environment Federation (1992) *Standard Methods for the Examination of Water and Wastewater.* 18th Ed., Washington, D.C., p. 4-66.

2. c. 0.1

   pH meters may have scales reading to the nearest 0.001 pH units under ideal conditions, but accurate and reproducible readings of 0.1 pH units represent the limit of accuracy under normal operating conditions.

   Refer to American Public Health Association; American Water Works Association; and Water Environment Federation (1992) *Standard Methods for the Examination of Water and Wastewater.* 18th Ed., Washington, D.C., p. 4-66.

3. a. Increases

   Neutral pH is a point of equilibrium between hydrogen ions and hydroxyl ions. The neutral point of pH is temperature dependent. As temperature increases, the neutral pH decreases (at 60 °C, neutral pH is 6.5) and, as temperature decreases, the neutral pH increases.

   Refer to American Public Health Association; American Water Works Association; and Water Environment Federation (1992) *Standard Methods for the Examination of Water and Wastewater.* 18th Ed., Washington, D.C., p. 4-65.

4. b. pH

Alkalinity and acidity are the acid- and base-neutralizing capacities, expressed as milligrams per liter $CaCO_3$.

Refer to American Public Health Association; American Water Works Association; and Water Environment Federation (1992) *Standard Methods for the Examination of Water and Wastewater.* 18th Ed., Washington, D.C., p. 4-65.

5. c. Lowers the pH

Refer to Water Environment Federation (1996) *Operation of Municipal Wastewater Treatment Plants.* 5th Ed., Manual of Practice No. 11, Alexandria, Va., p. 483.

6. c. The pH reading changes fewer than 0.1 pH units

Refer to U.S. Environmental Protection Agency (1983) *Methods for Chemical Analysis of Water and Wastewater.* EPA-600/4-79-020 (revised March 1983), Environ. Monit. Support Lab., Cincinnati, Ohio, p. 150.1-2.

7. c. Oily material and/or precipitated solids

Any material that coats the pH electrode or clogs the junction will cause interference. Also use caution in handling glass electrodes and avoid scratching the bulb. Refer to individual manufacturer instructions for care and maintenance of electrodes.

Refer to U.S. Environmental Protection Agency (1983) *Methods for Chemical Analysis of Water and Wastewater.* EPA-600/4-79-020 (revised March 1983), Environ. Monit. Support Lab., Cincinnati, Ohio, p. 150.1-1.

8. a. Daily

Working solutions should be replaced daily.

Refer to U.S. Environmental Protection Agency (1983) *Methods for Chemical Analysis of Water and Wastewater.* EPA-600/4-79-020 (revised March 1983), Environ. Monit. Support Lab., Cincinnati, Ohio, p. 150.1-2.

9. d. Leave it uncovered

This allows a free flow of electrolyte across the electrode junction which, if interrupted, causes an increase in response time and drift in the reading.

Refer to Orion Research, Inc. *Orion Ross® Sure-Flow Electrodes Instruction Manual.* 227355-001 Rev A, Beverly, Mass., p. 6.

10. d. None, analyze immediately

pH is not stable and will change over time.

Refer to U.S. Environmental Protection Agency (1983) *Methods for Chemical Analysis of Water and Wastewater.* EPA-600/4-79-020 (revised March 1983), Environ. Monit. Support Lab., Cincinnati, Ohio, p. xvi.

11. a. Avoid vigorous shaking or mixing of the sample

Agitation will increase levels of dissolved gases in a sample.

Refer to American Public Health Association; American Water Works Association; and Water Environment Federation (1992) *Standard Methods for the Examination of Water and Wastewater.* 18th Ed., Washington, D.C., p. 2-23.

12. c. 0.1 M $Na_2S_2O_3$

Sodium thiosulfate neutralizes chlorine residual in the sample.

Refer to American Public Health Association; American Water Works Association; and Water Environment Federation (1992) *Standard Methods for the Examination of Water and Wastewater.* 18th Ed., Washington, D.C., p. 2-24.

13. b. Phenolphthalein indicator endpoint of 8.3

Refer to American Public Health Association; American Water Works Association; and Water Environment Federation (1992) *Standard Methods for the Examination of Water and Wastewater.* 18th Ed., Washington, D.C., p. 2-23.

14. c. 10 to 1000 mg/L as $CaCO_3$

Refer to U.S. Environmental Protection Agency (1983) *Methods for Chemical Analysis of Water and Wastewater.* EPA-600/4-79-020 (revised March 1983), Environ. Monit. Support Lab., Cincinnati, Ohio, p. 305.1-1.

15. c. Slowly adding drops of titrant as the endpoint pH is approached
Stirring the sample faster is incorrect because this could increase the dissolved gases in the sample. Slowly titrating allows time for the titrant to fully react with the sample and for the pH electrode to respond. However, waiting 10 minutes between each titrant addition is excessive.

Refer to U.S. Environmental Protection Agency (1983) *Methods for Chemical Analysis of Water and Wastewater.* EPA-600/4-79-020 (revised March 1983), Environ. Monit. Support Lab., Cincinnati, Ohio, p. 305.1-1.

16. c. A strong base required to bring the sample to a designated pH
The titrant NaOH is a base and is used to neutralize the acid in the sample.

Refer to American Public Health Association; American Water Works Association; and Water Environment Federation (1992) *Standard Methods for the Examination of Water and Wastewater.* 18th Ed., Washington, D.C., p. 2-23.

17. c. Thoroughly but gently mixes the sample and titrant with no splashing

This allows the sample to remain mixed while preventing the loss or gain of CO_2 to or from the atmosphere.

Refer to American Public Health Association; American Water Works Association; and Water Environment Federation (1992) *Standard Methods for the Examination of Water and Wastewater.* 18th Ed., Washington, D.C., p. 2-25.

18. a. 0.02 N NaOH

Refer to American Public Health Association; American Water Works Association; and Water Environment Federation (1992) *Standard Methods for the Examination of Water and Wastewater.* 18th Ed., Washington, D.C., p. 2-24.

19. c. 14 days at 4 °C

Refer to American Public Health Association; American Water Works Association; and Water Environment Federation (1992) *Standard Methods for the Examination of Water and Wastewater.* 18th Ed., Washington, D.C., p. 1-22; and U.S. Environmental Protection Agency (1983) *Methods for Chemical Analysis of Water and Wastewater.* EPA-600/4-79-020 (revised March 1983), Environ. Monit. Support Lab., Cincinnati, Ohio, p. xvii.

20. a. Specific pH endpoint used

Refer to American Public Health Association; American Water Works Association; and Water Environment Federation (1992) *Standard Methods for the Examination of Water and Wastewater.* 18th Ed., Washington, D.C., p. 2-25.

21. c. 8.2

Refer to U.S. Environmental Protection Agency (1983) *Methods for Chemical Analysis of Water and Wastewater.* EPA-600/4-79-020 (revised March 1983), Environ. Monit. Support Lab., Cincinnati, Ohio, p. 305.1-2.

22. a. 3.7 and 8.3

Standard Methods refers to "methyl orange acidity" (pH 3.7) and "phenolphthalein" or total acidity (pH 8.3) as the typical means of determining the end points of the acidity titration.

Refer to American Public Health Association; American Water Works Association; and Water Environment Federation (1992) *Standard Methods for the Examination of Water and Wastewater.* 18th Ed., Washington, D.C., p. 2-23.

23. a. Acidity, mg/L = (Titrant Used, mL $\times$ Normality of NaOH Titrant $\times$ 50 000)/Sample Volume, mL

Refer to American Public Health Association; American Water Works Association; and Water Environment Federation (1992) *Standard Methods for the Examination of Water and Wastewater.* 18th Ed., Washington, D.C., p. 2-25.

24. d. milligrams per liter $CaCO_3$

Refer to American Public Health Association; American Water Works Association; and Water Environment Federation (1992) *Standard Methods for the Examination of Water and Wastewater.* 18th Ed., Washington, D.C., p. 2-27.

25. a. Alkalinity, mg/L = (Volume of Acid Titrant, mL × Normality of Acid Titrant × 50 000)/Sample Volume, mL

Refer to American Public Health Association; American Water Works Association; and Water Environment Federation (1992) *Standard Methods for the Examination of Water and Wastewater.* 18th Ed., Washington, D.C., p. 2-27.

26. c. 4.5

Refer to American Public Health Association; American Water Works Association; and Water Environment Federation (1992) *Standard Methods for the Examination of Water and Wastewater.* 18th Ed., Washington, D.C., p. 2-26.

27. c. Clean the electrodes occasionally between titrations

Refer to U.S. Environmental Protection Agency (1983) *Methods for Chemical Analysis of Water and Wastewater.* EPA-600/4-79-020 (revised March 1983), Environ. Monit. Support Lab., Cincinnati, Ohio, p. 150.1-1.

28. d. Having a titration volume greater than 50 mL
Generally, a 50-mL buret is used for such titrations, and accuracy is lost if the buret must be filled multiple times to complete the titration.

Refer to U.S. Environmental Protection Agency (1983) *Methods for Chemical Analysis of Water and Wastewater.* EPA-600/4-79-020 (revised March 1983), Environ. Monit. Support Lab., Cincinnati, Ohio, p. 310.1-2.

29. d. 25 °C ± 2 °C

Refer to U.S. Environmental Protection Agency (1983) *Methods for Chemical Analysis of Water and Wastewater.* EPA-600/4-79-020 (revised March 1983), Environ. Monit. Support Lab., Cincinnati, Ohio, p. 310.1-1.

30. a. Keep the air space above the solution to a minimum
Minimizing the air space above the sample reduces the absorption of CO_2 from the atmosphere.

Refer to U.S. Environmental Protection Agency (1983) *Methods for Chemical Analysis of Water and Wastewater.* EPA-600/4-79-020 (revised March 1983), Environ. Monit. Support Lab., Cincinnati, Ohio, p. 310.1-1.

31. b. 0.02 N H_2SO_4 or HCl for alkalinity concentrations < 1000 mg $CaCO_3$/L
The strength of the titrant selected is dependent on the anticipated range of the alkalinity in the sample.

Refer to U.S. Environmental Protection Agency (1983) *Methods for Chemical Analysis of Water and Wastewater.* EPA-600/4-79-020 (revised March 1983), Environ. Monit. Support Lab., Cincinnati, Ohio, p. 310.1-2.

32. c. 4.5

Refer to U.S. Environmental Protection Agency (1983) *Methods for Chemical Analysis of Water and Wastewater.* EPA-600/4-79-020 (revised March 1983), Environ. Monit. Support Lab., Cincinnati, Ohio p. 310.1-2.

33. a. Capacity to neutralize acids

Refer to Water Environment Federation (1996) *Operation of Municipal Wastewater Treatment Plants.* 5th Ed., Manual of Practice No. 11, Alexandria, Va., p. 483.

34. c. Alkalinity

Refer to American Public Health Association; American Water Works Association; and Water Environment Federation (1992) *Standard Methods for the Examination of Water and Wastewater.* 18th Ed., Washington, D.C., p. 2-27.

35. d. 100 times as acidic

Smith, R.-K. (1999) *Handbook of Environmental Analysis.* 4th Ed., Genium Publishing, Schenectady, N.Y., p. 231.

36. c. A grab sample

Refer to American Public Health Association; American Water Works Association; and Water Environment Federation (1992) *Standard Methods for the Examination of Water and Wastewater.* 18th Ed., Washington, D.C., p. 1-22.

37. b. Indicator

Refer to American Public Health Association; American Water Works Association; and Water Environment Federation (1992) *Standard Methods for the Examination of Water and Wastewater.* 18th Ed., Washington, D.C., p. 2-23.

38. c. 7 and 10

Refer to American Public Health Association; American Water Works Association; and Water Environment Federation (1992) *Standard Methods for the Examination of Water and Wastewater*. 18th Ed., Washington, D.C., p. 4-68.

39. a. Acidic

Smith, R.-K. (1999) *Handbook of Environmental Analysis*. 4th Ed., Genium Publishing, Schenectady, N.Y., p. 231.

40. d. pH values cannot be averaged

It cannot be determined from the data given because standard units cannot be averaged.

Smith, R.-K. (1999) *Handbook of Environmental Analysis*. 4th Ed., Genium Publishing, Schenectady, N.Y., p. 231.

41. d. Sodium carbonate (Na_2CO_3)

Refer to American Public Health Association; American Water Works Association; and Water Environment Federation (1992) *Standard Methods for the Examination of Water and Wastewater*. 18th Ed., Washington, D.C., p. 2-26.

42. a. The negative logarithm of the hydrogen ion activity

Refer to American Public Health Association; American Water Works Association; and Water Environment Federation (1992) *Standard Methods for the Examination of Water and Wastewater*. 18th Ed., Washington, D.C., p. 4-65.

43. a. Septic anaerobic conditions in the collection system
Acid-forming organisms convert organic compounds into volatile organic acids.

Refer to Water Environment Federation (1996) *Operation of Municipal Wastewater Treatment Plants.* 5th Ed., Manual of Practice No. 11, Alexandria, Va., p. 972.

44. b. Carbonate, hydroxide, and bicarbonate

Refer to American Public Health Association; American Water Works Association; and Water Environment Federation (1992) *Standard Methods for the Examination of Water and Wastewater.* 18th Ed., Washington, D.C., p. 2-27.

45. a. Bicarbonate alkalinity

Refer to American Public Health Association; American Water Works Association; and Water Environment Federation (1992) *Standard Methods for the Examination of Water and Wastewater.* 18th Ed., Washington, D.C., p. 2-28.

46. d. Temperature

Refer to American Public Health Association; American Water Works Association; and Water Environment Federation (1992) *Standard Methods for the Examination of Water and Wastewater.* 18th Ed., Washington, D.C., p. 4-66.

47. c. 3

Refer to American Public Health Association; American Water Works Association; and Water Environment Federation (1992) *Standard Methods for the Examination of Water and Wastewater.* 18th Ed., Washington, D.C., p. 4-68.

48. b. Daily

Refer to American Public Health Association; American Water Works Association; and Water Environment Federation (1992) *Standard Methods for the Examination of Water and Wastewater*. 18th Ed., Washington, D.C., p. 1-5.

49. c. Temperature

Refer to American Public Health Association; American Water Works Association; and Water Environment Federation (1992) *Standard Methods for the Examination of Water and Wastewater*. 18th Ed., Washington, D.C., p. 4-66.

50. c. pH

Refer to American Public Health Association; American Water Works Association; and Water Environment Federation (1992) *Standard Methods for the Examination of Water and Wastewater*. 18th Ed., Washington, D.C., p. 4-65.

51. a. 4.5

Refer to American Public Health Association; American Water Works Association; and Water Environment Federation (1992) *Standard Methods for the Examination of Water and Wastewater*. 18th Ed., Washington, D.C., p. 2-26.

BIOCHEMICAL OXYGEN DEMAND/CARBONACEOUS BIOCHEMICAL OXYGEN DEMAND/CHEMICAL OXYGEN DEMAND/DISSOLVED OXYGEN

Questions

1. A DO sample should not be agitated or remain in contact with the atmosphere because

    a. The solids may not settle out
    b. The gaseous content of the sample may change
    c. The acidity in the sample will reduce the dissolved oxygen
    d. The sample may give off toxic fumes

2. For *Standard Methods* 4500-O B. [Oxygen (Dissolved): Iodometric Methods], why should the sample bottle overflow two or three times when collecting a sample under pressure?

    a. To get enough sample
    b. To get a full bottle
    c. To prevent the entrainment of air bubbles
    d. To remove excess pollutants

3. What methods are used for measuring the DO content of a sample?

    a. The Winkler (or iodometric) method and its modifications and the electrometric (membrane electrode) method
    b. The oxygen phosphate method and the electrometric (membrane electrode) method
    c. The nitroferrous electrode and azide modification
    d. Infrared and MnO_4 modification

4. What method of DO testing is best for field testing wastewater?

    a. Alum flocculation modification
    b. Iodometric method
    c. Membrane electrode (electrometric) method
    d. Azide modification

5. What is the most important parameter affecting biological activity in a wastewater treatment process?

    a. Grease in the influent
    b. DO
    c. SS
    d. $FeCl_2$

6. What does DO refer to?

    a. Dissolved organics
    b. Dissolved oxygen
    c. Direct oxidation
    d. Diamino oxide

7. What is the primary advantage of using a DO probe and meter?

    a. Measurements can be made directly in the waste stream
    b. Less care is needed in sampling
    c. The probe and meter need little or no preparation before using
    d. Samples can be fixed and analyzed at the operator's convenience

8. Which modification of the Winkler test is used for biological flocs with high oxygen utilization rates?

   a. Alum flocculation
   b. Azide
   c. $CuSO_4$–sulfamic acid flocculation
   d. MnO_4

9. The temperature of the incubator used for a BOD test should be maintained at

   a. 20 °F
   b. 35 °F
   c. 20 °C
   d. 68 °C

10. What test indicates the rate of oxidation and provides an indirect estimate of the concentration of the waste?

    a. BOD
    b. TSS
    c. DO
    d. TOC

11. What is the basic formula used to determine BOD?

    a. Oxygen depletion of the blank minus oxygen depletion of the sample divided by the percent of sample
    b. Oxygen depletion during incubation divided by the dilution factor of the incubated sample
    c. DO of the raw sample multiplied by the DO of the incubated sample
    d. DO of the dilution water divided by the percent of sample

12. If there is DO present in a BOD sample, what color will the floc turn after the alkaline iodide–sodium azide solution is added?

    a. Blue
    b. Brown
    c. White
    d. Purple

13. The oxygen depleted during the 5-day BOD test is used by which of the following?

    a. Chemicals
    b. Microorganisms
    c. Industrial discharges
    d. Ammonia

14. The equation for calculating BOD in milligrams per liter is

    a. $\dfrac{\text{Initial DO, mg/L} - \text{Final DO, mg/L}}{\text{Initial DO, mg/L}} \times 100\%$

    b. $\dfrac{\dfrac{(\text{Initial DO, mg/L} - \text{Final DO, mg/L})}{\text{Sample Volume, mL}}}{\text{Bottle Volume, mL}}$

    c. $\dfrac{(\text{Initial DO, mg/L} - \text{Final DO, mg/L})(\text{Sample Volume, mL})}{\text{Bottle Volume, mL}}$

    d. $\dfrac{\text{Initial DO, mg/L} - \text{Final DO, mg/L}}{\text{Time, minutes}}$

15. A DO sample to be analyzed by the Winkler method, which has been acidified in the field, may be held for

    a. No time, must be analyzed immediately

b. 0.5 hour
c. 6 hours
d. 8 hours

16. To prepare a zero DO standard for the membrane electrode test

    a. Add excess Na_2SO_3 and a trace of $CoCl_2$
    b. Add $MnSO_4$ and shake well
    c. Add alkali iodide azide and let stand for 20 minutes
    d. Add excess NaCl

17. In determining COD by the dichromate reflux method, the endpoint of the titration is reached when

    a. Excess ferric ions complex the methyl orange indicator
    b. Ferrous ions are no longer oxidized by the residual dichromate and the presence of phenanthroline forms a red complex
    c. Ferrous ammonium sulfate is completely oxidized
    d. Ferrous ammonium sulfate is completely neutralized

18. After 5 days incubation, a BOD sample should have a DO depletion of at least

    a. 0.5 mg/L
    b. 1.0 mg/L
    c. 2.0 mg/L
    d. 13.5 mg/L

19. The Winkler method is used for determining the level of which of the following?

    a. Amount of DO
    b. Fecal coliform levels

c. Nonfilterable residue
d. Free ammonia

20. BOD is the

a. The measure of the biological diversity in a sample under standardized conditions
b. Amount of carbonaceous matter present in a representative sample under standardized conditions
c. The measure of the oxygen required during the stabilization of decomposable matter by aerobic bacterial action under standardized conditions
d. Ratio of the available DO to the oxygen demand under normal temperature and pressure

21. What is the proper incubator temperature setting for BOD analysis?

a. 20 °F ± 1 °F
b. 68 °F ± 1 °F
c. 4 °C ± 1 °C
d. 20 °C ± 1 °C

22. In the CBOD test, 2-chloro-6-(trichlormethyl)pyridine is used to suppress

a. Nitrification
b. Carbonation
c. Gasification
d. Denitrification

23. Seeding BOD dilution water with a microbial population is necessary when analyzing

    a. Influent wastewater
    b. Receiving streams
    c. Process effluent (e.g., primary clarifier effluent)
    d. Chlorinated effluent

24. The water quality for BOD dilution and reagent water should be

    a. Type 1 (>10 mΩ and relatively bacteria free)
    b. Type 2 (>1 mΩ and <1000 CFU/mL)
    c. Type 3 (0.1 mΩ and bacterial content is unimportant)
    d. Type 4 (chlorinated and filtered)

25. Which of the following solutions is NOT added to BOD dilution water?

    a. Na_2SO_3
    b. $MgSO_4$
    c. $CaCl_2$
    d. $FeCl_3$

26. The DO uptake of a seed control should never exceed

    a. 0.4 mg/L
    b. 0.6 mg/L
    c. 0.8 mg/l
    d. 1.0 mg/L

27. If a BOD sample is too acidic or alkaline, you should

    a. Adjust the dilution water pH to 6.5–7.5
    b. Adjust the seed pH to 6.5–7.5
    c. Adjust the sample pH to 6.5–7.5
    d. Adjust the DO level of the sample before processing

28. When adjusting the pH of a BOD sample, what normality of NaOH or H_2SO_4 solution should you use?

    a. 0.5 N
    b. 1.0 N
    c. 2.0 N
    d. Pure sodium hydroxide or sulfuric acid

29. If a BOD sample contains more than 9.0 mg/L of DO at 20 °C, you should

    a. Warm the sample above 20 °C
    b. Cool the sample below 20 °C
    c. Agitate or aerate the sample at 20 °C
    d. Proceed directly with the test procedure

30. Which sugar compound is used to prepare a standard for a BOD test?

    a. Sucrose
    b. Fructose
    c. Glucose
    d. Maltose

31. Which of the following GGA (BOD standard) measurements is the most accurate?

    a. 138 mg/L
    b. 168 mg/L
    c. 198 mg/L
    d. 228 mg/L

32. The difference between a CBOD and BOD test is

    a. Incubation period
    b. Volume of sample used
    c. Incubation temperature above 20 °C
    d. Addition of nitrification inhibitor

33. The "C" in CBOD stands for

    a. Carbonaceous
    b. Chemical
    c. Combined
    d. Compound

34. Which of the following chemicals is NOT used in a COD test?

    a. $HgSO_4$
    b. $MgSO_4$
    c. Ferroin indicator
    d. $AgSO_4$

35. Given the following information, determine the COD of the sample

   Amount of FAS in the blank = 5 mL
   Amount of FAS titrant used for the sample = 4.4 mL
   Molarity of FAS titrant = 0.101 M
   Amount of sample = 2.5 mL

   a. 162 mg O_2/L
   b. 194 mg O_2/L
   c. 226 mg O_2/L
   d. 258 mg O_2/L

36. Which of the following phrases represents the correlation between the COD and BOD values of a specific wastewater sample?

   a. COD is larger than BOD
   b. COD is smaller than BOD
   c. COD is equal to BOD
   d. COD is unrelated to BOD

37. The BOD:COD for your facility's influent is 0.6. If the COD of your influent is 360 mg/L, what is the approximate BOD of your influent?

   a. 20 mg/L
   b. 200 mg/L
   c. 400 mg/L
   d. 600 mg/L

38. To ensure the most accurate results for BOD analysis, you should

   a. Preserve with H_2SO_4

b. Store sample at 20 °C

c. Preserve with HCl

d. Test immediately on site

39. What is the minimum DO uptake that will produce the most reliable results for the BOD test?

a. 0.2 mg/L
b. 1.0 mg/L
c. 2.0 mg/L
d. 2.5 mg/L

40. What is wrong with a COD sample that turns green during the reflux?

a. COD is below the detection limit
b. $K_2Cr_2O_7$ solution was not standardized
c. Too much FAS was added to the sample
d. All of the oxidizer has been depleted

Answers

1. b. The gaseous content of the sample may change

    Agitation or exposure to the atmosphere may add more oxygen to the sample than would normally be there.

    Refer to American Public Health Association; American Water Works Association; and Water Environment Federation (1992) *Standard Methods for the Examination of Water and Wastewater.* 18th Ed., Washington, D.C., p. 4-98.

2. c. To prevent the entrainment of air bubbles

    Fill the bottle to overflowing and prevent turbulence and formation of bubbles while filling.

    Refer to American Public Health Association; American Water Works Association; and Water Environment Federation (1992) *Standard Methods for the Examination of Water and Wastewater.* 18th Ed., Washington, D.C., p. 4-99.

3. a. The Winkler (or iodometric) method and its modifications and the electrometric (membrane electrode) method

    The iodometric method is a titrimetric procedure based on the oxidizing property of DO, and the electrometric (membrane electrode) method is based on the rate of diffusion of molecular oxygen across a membrane. The choice of procedure depends on interferences present; the accuracy desired; and, in some cases, convenience or expedience.

    Refer to American Public Health Association; American Water Works Association; and Water Environment Federation (1992) *Standard Methods for the Examination of Water and Wastewater.* 18th Ed., Washington, D.C., p. 4-98.

4. c. Membrane electrode (electrometric) method

The membrane electrode method is used for field testing because it is subject to fewer interferences and it can be adapted for continuous monitoring. Although various modifications of the iodometric method have been developed to eliminate or minimize effects of interferences, extra laboratory equipment and chemicals are needed for these modifications.

Refer to American Public Health Association; American Water Works Association; and Water Environment Federation (1992) *Standard Methods for the Examination of Water and Wastewater.* 18th Ed., Washington, D.C., p. 4-103.

5. b. DO

DO stands for dissolved oxygen. All phases of microbial growth require sufficient air (oxygen) to meet the demands of the process.

Refer to Water Environment Federation (1996) *Operation of Municipal Wastewater Treatment Plants.* 5th Ed., Manual of Practice No. 11, Alexandria, Va., p. 640.

6. b. Dissolved oxygen

Refer to American Public Health Association; American Water Works Association; and Water Environment Federation (1992) *Standard Methods for the Examination of Water and Wastewater.* 18th Ed., Washington, D.C., p. 4-98.

7. a. Measurements can be made directly in the waste stream

Refer to American Public Health Association; American Water Works Association; and Water Environment Federation (1992) *Standard Methods for the Examination of Water and Wastewater.* 18th Ed., Washington, D.C., p. 4-103.

8. c. CuSO$_4$–sulfamic acid flocculation

Refer to American Public Health Association; American Water Works Association; and Water Environment Federation (1992) *Standard Methods for the Examination of Water and Wastewater*. 18th Ed., Washington, D.C., p. 4-103.

9. c. 20 °C

Refer to American Public Health Association; American Water Works Association; and Water Environment Federation (1992) *Standard Methods for the Examination of Water and Wastewater*. 18th Ed., Washington, D.C., p. 5-3.

10. a. BOD

The BOD, or biochemical oxygen demand, is an empirical test that determines the relative oxygen requirements of a water/wastewater sample under standardized conditions.

Refer to American Public Health Association; American Water Works Association; and Water Environment Federation (1992) *Standard Methods for the Examination of Water and Wastewater*. 18th Ed., Washington, D.C., p. 5-1.

11. b. Oxygen depletion during incubation divided by the dilution factor of the incubated sample

Refer to American Public Health Association; American Water Works Association; and Water Environment Federation (1992) *Standard Methods for the Examination of Water and Wastewater*. 18th Ed., Washington, D.C., p. 5-5.

12. b.   Brown

Refer to California State University (1992) *Operation of Municipal Wastewater Treatment Plants*. 4th Ed., Sacramento, Calif., p. 548.

13. b.   Microorganisms

The BOD test is conducted for 5 days at a controlled temperature in order to measure the amount of oxygen consumed by microorganisms as they break down the organic materials (pollutants) in wastewater.

Refer to American Public Health Association; American Water Works Association; and Water Environment Federation (1992) *Standard Methods for the Examination of Water and Wastewater*. 18th Ed., Washington, D.C., p. 5-1.

14. b. $$\frac{(\text{Initial DO, mg/L} - \text{Final DO, mg/L})}{\frac{\text{Sample Volume, mL}}{\text{Bottle Volume, mL}}}$$

Refer to American Public Health Association; American Water Works Association; and Water Environment Federation (1992) *Standard Methods for the Examination of Water and Wastewater*. 18th Ed., Washington, D.C., p. 5-5.

15. d.   8 hours

Refer to American Public Health Association; American Water Works Association; and Water Environment Federation (1992) *Standard Methods for the Examination of Water and Wastewater*. 18th Ed., Washington, D.C., pp. 1-22 and 4-99.

16. a. Add excess Na$_2$SO$_3$ and a trace of CoCl$_2$

Refer to American Public Health Association; American Water Works Association; and Water Environment Federation (1992) *Standard Methods for the Examination of Water and Wastewater.* 18th Ed., Washington, D.C., p. 4-104.

17. b. Ferrous ions are no longer oxidized by the residual dichromate and the presence of phenanthroline forms a red complex

Take the titration endpoint to be the first sharp color change from blue-green to reddish brown. The blue-green may reappear.

Refer to American Public Health Association; American Water Works Association; and Water Environment Federation (1992) *Standard Methods for the Examination of Water and Wastewater.* 18th Ed., Washington, D.C., pp. 5-7 and 5-8.

18. c. 2.0 mg/L

Refer to American Public Health Association; American Water Works Association; and Water Environment Federation (1992) *Standard Methods for the Examination of Water and Wastewater.* 18th Ed., Washington, D.C., p. 5-5.

19. a. Amount of DO

Refer to California Water Environment Association (1993) *Laboratory Analysts Study Manual.* Oakland, Calif., p. 6-3; and Smith, R.-K. (1999) *Lectures on Wastewater Analysis and Interpretation.* Genium Publishing, Schenectady, N.Y., p. 128.

20. c. The measure of the oxygen required during the stabilization of decomposable matter by aerobic bacterial action under standardized conditions

Refer to California Water Environment Association (1993) *Laboratory Analysts Study Manual.* Oakland, Calif., p. 6-4; and Smith, R.-K. (1999) *Lectures on Wastewater Analysis and Interpretation.* Genium Publishing, Schenectady, N.Y., p. 466.

21. d. 20 °C ± 1 °C

Refer to American Public Health Association; American Water Works Association; and Water Environment Federation (1992) *Standard Methods for the Examination of Water and Wastewater.* 18th Ed., Washington, D.C., p. 5-3.

22. a. Nitrification

Refer to American Public Health Association; American Water Works Association; and Water Environment Federation (1992) *Standard Methods for the Examination of Water and Wastewater.* 18th Ed., Washington, D.C., pp. 5-3 and 5-5.

23. d. Chlorinated effluent

Refer to American Public Health Association; American Water Works Association; and Water Environment Federation (1992) *Standard Methods for the Examination of Water and Wastewater.* 18th Ed., Washington, D.C., p. 5-4.

24. a. Type 1 (>10 mΩ and relatively bacteria free)
Type 1 reagent water is satisfactory in most cases. However, additional purification may be needed for certain methods such as total organic halide

(TOX) and trihalomethane formation potential (THMFP).

Refer to American Public Health Association; American Water Works Association; and Water Environment Federation (1992) *Standard Methods for the Examination of Water and Wastewater.* 18th Ed., Washington, D.C., pp. 1-32 and 5-1.

25. a. Na_2SO_3

Refer to American Public Health Association; American Water Works Association; and Water Environment Federation (1992) *Standard Methods for the Examination of Water and Wastewater.* 18th Ed., Washington, D.C., p. 5-3.

26. d. 1.0 mg/L

Refer to American Public Health Association; American Water Works Association; and Water Environment Federation (1992) *Standard Methods for the Examination of Water and Wastewater.* 18th Ed., Washington, D.C., p. 5-4.

27. c. Adjust the sample pH to 6.5–7.5

Refer to American Public Health Association; American Water Works Association; and Water Environment Federation (1992) *Standard Methods for the Examination of Water and Wastewater.* 18th Ed., Washington, D.C., p. 5-4.

28. b. 1.0 N

Refer to American Public Health Association; American Water Works Association; and Water Environment Federation (1992) *Standard Methods for the Examination of Water and Wastewater.* 18th Ed., Washington, D.C., p. 5-3.

29. c. Agitate or aerate the sample at 20 °C

Samples are considered to be supersaturated with DO if they contain more than 9 mg DO/L at 20 °C. In order to prevent oxygen loss during incubation, reduce the DO level to saturation at 20 °C by agitating, by vigorous shaking, or by aerating with clean, filtered compressed air.

Refer to American Public Health Association; American Water Works Association; and Water Environment Federation (1992) *Standard Methods for the Examination of Water and Wastewater.* 18th Ed., Washington, D.C., p. 5-4.

30. c. Glucose

Refer to American Public Health Association; American Water Works Association; and Water Environment Federation (1992) *Standard Methods for the Examination of Water and Wastewater.* 18th Ed., Washington, D.C., p. 5-3.

31. c. 198 mg/L

Refer to American Public Health Association; American Water Works Association; and Water Environment Federation (1992) *Standard Methods for the Examination of Water and Wastewater.* 18th Ed., Washington, D.C., p. 5-5.

32. d. Addition of nitrification inhibitor

Refer to American Public Health Association; American Water Works Association; and Water Environment Federation (1992) *Standard Methods for the Examination of Water and Wastewater.* 18th Ed., Washington, D.C., pp. 5-2 and 5-5.

33. a. Carbonaceous

Refer to American Public Health Association; American Water Works Association; and Water Environment Federation (1992) *Standard Methods for the Examination of Water and Wastewater.* 18th Ed., Washington, D.C., p. 5-2.

34. d. $AgSO_4$

Refer to American Public Health Association; American Water Works Association; and Water Environment Federation (1992) *Standard Methods for the Examination of Water and Wastewater.* 18th Ed., Washington, D.C., p. 5-9.

35. b. 194 mg O_2/L

Refer to American Public Health Association; American Water Works Association; and Water Environment Federation (1992) *Standard Methods for the Examination of Water and Wastewater.* 18th Ed., Washington, D.C., p. 5-9.

36. a. COD is larger than BOD

Refer to American Public Health Association; American Water Works Association; and Water Environment Federation (1992) *Standard Methods for the Examination of Water and Wastewater.* 18th Ed., Washington, D.C., p. 5-6.

37. b. 200 mg/L

BOD:COD = BOD/COD = 0.6

$$\frac{BOD}{360 \text{ mg/L}} = 0.6$$

BOD = 0.6 × 360 mg/L

BOD = 216 mg/L (rounded to 200 mg/L)

Refer to Water Environment Federation (2000) *Operations Training CD-ROM Series/Activated Sludge Process Control* [CD-ROM]. Alexandria, Va., Unit 4.

38. d. Test immediately on site

Refer to American Public Health Association; American Water Works Association; and Water Environment Federation (1992) *Standard Methods for the Examination of Water and Wastewater.* 18th Ed., Washington, D.C., p. 5-1.

39. c. 2.0 mg/L

Refer to American Public Health Association; American Water Works Association; and Water Environment Federation (1992) *Standard Methods for the Examination of Water and Wastewater.* 18th Ed., Washington, D.C., p. 5-4.

40. d. All of the oxidizer has been depleted

Refer to California State University (1992) *Operation of Municipal Wastewater Treatment Plants.* 4th Ed., Sacramento, Calif., p. 521.

CHLORINE/OZONE

Questions

1. An amperometric titration is often used when measuring

    a. Chloride
    b. Residual chlorine
    c. Alkalinity
    d. DO

2. The most important reason for chlorinating effluent is

    a. Decreasing BOD levels and controlling odors
    b. Improving scum and grease removal and chemical coagulation
    c. Disinfection
    d. Decreasing SS levels

3. When wastewater samples are chlorinated and the pH is lowered

    a. Nitrates are formed
    b. Chlorine gas is formed
    c. HClO is formed
    d. A precipitate is formed

4. Which of the following constituents is the best disinfecting agent?

    a. HClO
    b. Hypochlorite ion
    c. Monochloramine
    d. Dichloramine

5. Which one of the following is most likely to interfere with the DPD test?

   a. Amines
   b. Nitrite
   c. Oxidized ion
   d. Oxidized manganese

6. How is N, N-diethyl-p-phenylenediamine more commonly represented?

   a. NND
   b. SST
   c. DPD
   d. DPP

7. The chlorine residual may be determined using the reagent

   a. DPD
   b. EDTA
   c. PCB
   d. $Na_2S_2O_3$

8. The quantity of chlorine that is converted to inert or less active forms of chlorine by substances in wastewater is

   a. Free available chlorine
   b. Free chlorine residual
   c. Hypochlorite
   d. Total combined chlorine

9. Before you perform an amperometic titration for total residual chlorine, what two solutions should be added to the sample?

   a. Phenylarsine oxide and acetate buffer
   b. KI and phosphate buffer
   c. KI and phenylarsine oxide
   d. KI and acetate buffer

10. When performing an amperometric titration for chlorine, how many milliliters of sample should be analyzed?

    a. 100 mL
    b. 200 mL
    c. 500 mL
    d. 1000 mL (1L)

11. When using a spectrophotometer to measure chlorine, what wave length of light should be used?

    a. 460 nm
    b. 515 nm
    c. 540 nm
    d. 630 nm

12. An effluent with a high level of ammonia nitrogen has chlorine residual that is mostly

    a. Free chlorine
    b. Total chlorine
    c. Amperometric chlorine
    d. Combined chlorine

Answers

1. b. Residual chlorine

   Refer to American Public Health Association; American Water Works Association; and Water Environment Federation (1992) *Standard Methods for the Examination of Water and Wastewater.* 18th Ed., Washington, D.C., p. 4-36.

2. c. Disinfection
   Chlorination serves primarily to destroy or deactivate disease-producing microorganisms.

   Refer to American Public Health Association; American Water Works Association; and Water Environment Federation (1992) *Standard Methods for the Examination of Water and Wastewater.* 18th Ed., Washington, D.C., p. 4-36.

3. c. HCIO is formed

   Refer to American Public Health Association; American Water Works Association; and Water Environment Federation (1992) *Standard Methods for the Examination of Water and Wastewater.* 18th Ed., Washington, D.C., p. 4-36.

4. a. HCIO

   Refer to California State University (1992) *Operation of Municipal Wastewater Treatment Plants.* 4th Ed., Sacramento, Calif., p. 344.

5. d. Oxidized manganese

    References: American Public Health Association; American Water Works Association; and Water Environment Federation (1992) *Standard Methods for the Examination of Water and Wastewater.* 18th Ed., Washington, D.C., p. 4-44.

6. c. DPD

    Refer to American Public Health Association; American Water Works Association; and Water Environment Federation (1992) *Standard Methods for the Examination of Water and Wastewater.* 18th Ed., Washington, D.C., p. 4-43.

7. a. DPD

    Refer to American Public Health Association; American Water Works Association; and Water Environment Federation (1992) *Standard Methods for the Examination of Water and Wastewater.* 18th Ed., Washington, D.C., p. 4-43.

8. d. Total combined chlorine

    Refer to American Public Health Association; American Water Works Association; and Water Environment Federation (1992) *Standard Methods for the Examination of Water and Wastewater.* 18th Ed., Washington, D.C., p. 4-36.

9. d. KI and acetate buffer

    Refer to American Public Health Association; American Water Works Association; and Water Environment Federation (1992) *Standard Methods for the Examination of Water and Wastewater.* 18th Ed., Washington, D.C., p. 4-42.

10. b. 200 mL

Refer to American Public Health Association; American Water Works Association; and Water Environment Federation (1992) *Standard Methods for the Examination of Water and Wastewater.* 18th Ed., Washington, D.C., p. 4-42.

11. b. 515 nm

Refer to American Public Health Association; American Water Works Association; and Water Environment Federation (1992) *Standard Methods for the Examination of Water and Wastewater.* 18th Ed., Washington, D.C., p. 4-46.

12. d. Combined chlorine

Refer to American Public Health Association; American Water Works Association; and Water Environment Federation (1992) *Standard Methods for the Examination of Water and Wastewater.* 18th Ed., Washington, D.C., p. 4-36.

MICROBIOLOGICAL EXAMINATION

Questions

1. The density of coliform bacteria estimated using the multiple-tube fermentation technique is expressed as

    a. DPD
    b. MF/mL
    c. mgd
    d. MPN/100 mL

2. When collecting samples for bacteriological examination, $Na_2S_2O_3$ is used to

    a. Buffer the pH to 7.0
    b. Buffer the pH to 8.5
    c. Oxidize the combined chlorine
    d. Neutralize any remaining disinfectant

3. A negative control culture for fecal coliform is

    a. *Enterobacter aerogenes*
    b. *Escherichia coli*
    c. *Cryptosporidium parvum*
    d. *Giardia lamblia*

4. When performing microbiological examinations, sterilize glassware by placing it in an oven at

    a. 103 °C for 1 hour or more
    b. 121 °C for 15 minutes

c. 170 °C for at least 1 hour
d. 550 °C for 20 minutes

5. The membrane filter procedure for fecal coliform requires a specific incubation temperature of

a. 44.5 °C ± 0.1 °C
b. 45.4 °C ± 0.1 °C
c. 44.5 °C ± 0.2 °C
d. 45.4 °C ± 0.5 °C

6. When computing the density of a fecal coliform plate, what is the desired range of colonies?

a. 1 to 100 CFU/100 mL
b. 10 to 50 CFU/100 mL
c. 20 to 60 CFU/100 mL
d. 20 to 100 CFU/100 mL

7. When using the autoclave to sterilize glassware used for determining fecal coliform, the autoclave must reach a temperature of

a. 100 °C
b. 121 °C
c. 250 °C
d. 270 °C

8. Which of the following is NOT characteristic of the fecal coliform group of bacteria?

    a. Non-spore-forming
    b. Rod-shaped
    c. Gram-positive
    d. Gram-negative

9. Fecal coliform must have certain characteristics to be a pathogen indicator. Which of the following is NOT a characteristic?

    a. Should be present in greater numbers than the pathogen
    b. Should be a pathogen
    c. Should be found in mammals
    d. Should not be a pathogen

9. The MPN test for coliform bacteria has three test phases. Which of the following is NOT one of these phases?

    a. Presumptive
    b. Completed
    c. Calculated
    d. Confirmed

11. Which of these groups of materials are used to test for coliform bacteria by the MPN method?

    a. Brilliant green bile broth, refrigerator, and membrane filters
    b. Lauryl tryptose broth, incubator, and autoclave
    c. M-endo broth, autoclave, thermometer, and fermentation tubes
    d. Petri dishes, blood agar, and sterile graduated cylinders

12. What is the benefit of using toxicity testing for water pollution evaluation?

    a. Toxicity testing is cost efficient
    b. Chemical and physical tests do not assess the effect on aquatic biota
    c. Toxicity tests are longer tests, so they are better
    d. Chemical and physical tests do not tell you anything

13. To determine if a toxicity effect is likely to be observed, a single concentration is set up with multiple replicates for 24 to 96 hours. This is a

    a. Range-finding test
    b. Screening test
    c. Definitive test
    d. Dilution threshold test

14. What toxicity test would typically be used for a screening test and have an LC_{50} as the single endpoint?

    a. Acute test
    b. Chronic test
    c. Lethal dose concentration
    d. Total toxic density

15. Which test would assess toxic effects on different life stages and/or growth and reproduction of an organism?

    a. Acute test
    b. Chronic test
    c. Lowest observed effect test
    d. No observed effect test

16. Which toxicity test is most challenging for use on samples with high BOD and/or COD or high bacterial populations?

    a. Static test
    b. Renewal test
    c. Flow-through test
    d. Acute test

17. The toxicant concentration estimated to produce death in a specified portion of test organisms is the

    a. LC
    b. EC
    c. IC
    d. NOEC

18. The toxicant concentration estimated to cause a specific percentage inhibition or impairment in a qualitative biological function is the

    a. LC
    b. EC
    c. IC
    d. NOEC

19. The amount of toxicant that enters the organism is called the

    a. Dose
    b. Concentration
    c. Effluent
    d. Absorbent

20. Which of the following duplicates all of the conditions of a toxicity test but contains no test materials?

    a. Static
    b. Number 1
    c. Control
    d. First

21. The toxicity test in which organisms are exposed to solutions of the same composition, which are replaced periodically during the test period, is called the

    a. Flow-through test
    b. Renewal test
    c. Static test
    d. Periodic test

22. The lowest dose at which a measured response is statistically significantly different from that for the control is the

    a. LC
    b. EC
    c. IC
    d. LOEC

23. What does the 96-hour LC_{50} toxicity test demonstrate?

    a. Short-term toxicity
    b. Long-term toxicity
    c. The degree of organic loading
    d. The effect of temperature on toxicity

24. Why are different test species used in toxicity testing?

    a. It makes the testing more cost effective
    b. It helps to determine the effect of acidity on toxicity
    c. U.S. EPA protocol requires that at least five different species be used
    d. Different species have various levels of sensitivity to different toxicants

25. *Pimephales promelas*, which may be used in acute and/or chronic toxicity tests, are more commonly known as

    a. Zooplankton
    b. Fathead minnows
    c. Speckled trout
    d. Water fleas

26. What concentration of $Na_2S_2O_3$ solution is used to dechlorinate samples of wastewater effluent collected for microbiological testing?

    a. 3%
    b. 5%
    c. 10%
    d. 20%

27. Which of the following is added to 1 L of reagent-grade water to prepare buffered dilution water for a bacteriological examination?

    a. 1.25 mL KH_2PO_4 buffer and 5 mL $MgCl_2$ solution
    b. 5 mL KH_2PO_4 buffer and 1.25 mL $MgCl_2$ solution
    c. 50 mL KH_2PO_4 buffer and 5 mL $MgCl_2$ solution
    d. 34 mL KH_2PO_4 buffer and 50 mL $MgCl_2$ solution

28. Which of the following techniques should be used if you are asked to perform a 1:10 (0.1-m) dilution?

    a. Place 10 mL of sample into 99 mL of blank
    b. Run 0.1 mL of sample across a filter
    c. Place 1 mL of sample in 99 mL of blank and run 10 mL of inoculated blank across a filter
    d. Place 1 mL of sample in 999 mL of blank

29. For all transfers in a dilution series, you should use

    a. The same pipet for each transfer
    b. Two pipets: one for initial transfer and another for all subsequent transfers
    c. Separate sterile pipets for each transfer
    d. A sterile graduated cylinder

30. What is the minimum number of samples required for a 40 CFR Part 503 fecal coliform (pathogen reduction) test?

    a. 1
    b. 3
    c. 5
    d. 7

31. In what units are the results of the 40 CFR Part 503 pathogen reduction test expressed?

    a. CFU/g
    b. MPG
    c. CFU
    d. a. or b.

32. What is the minimum recommended magnification for microscopic examination of microbial populations in mixed liquor or other biomass?

    a. 10X
    b. 50X
    c. 100X
    d. 1000X

33. The color of fecal coliform colonies cultured on M-FC media is

    a. Blue
    b. Orange
    c. Gold–green (sheen)
    d. Gray

34. The color of total coliform colonies cultured on M-endo media is

    a. Blue
    b. Orange
    c. Gold–green (sheen)
    d. Gray

35. What is the desired density of total coliform colonies per membrane using the membrane filtration procedure?

    a. 0 to 50
    b. 10 to 100
    c. 20 to 60
    d. 20 to 80

36. The desired range for fecal coliform colonies using the membrane filtration technique is

    a. 0 to 50
    b. 1 to 10
    c. 20 to 60
    d. 20 to 80

37. In a fermentation tube containing lauryl tryptose broth or brilliant green bile, which observation confirms the presence of coliform bacteria?

    a. Color change
    b. Gas bubbles
    c. Clumps
    d. A 50% increase in volume

38. Before disposal of biologically contaminated materials such as cultures, media plates, and samples, what should be done to them?

    a. Wash
    b. Freeze
    c. Sterilize
    d. Cover

39. Diseases are most likely to be caused by

    a. Viruses
    b. Bacteria
    c. Pathogens
    d. Parasites

40. The purpose of adding $Na_2S_2O_3$ to a sample for coliform analysis is to

    a. Ensure sterilization of the bottle
    b. Remove any residual chlorine
    c. Fix the sample until it can be analyzed in the laboratory
    d. Extend the holding time to 24 hours

41. The incubation temperature used in the membrane filter technique for measuring levels of fecal coliform is

    a. 20 °C
    b. 35 °C
    c. 44.5 °C
    d. 104 °C

Answers

1. d.   MPN/100 mL

    Refer to American Public Health Association; American Water Works Association; and Water Environment Federation (1992) *Standard Methods for the Examination of Water and Wastewater*. 18th Ed., Washington, D.C., p. 9-49.

2. d.   Neutralize any remaining disinfectant

    Refer to American Public Health Association; American Water Works Association; and Water Environment Federation (1992) *Standard Methods for the Examination of Water and Wastewater*. 18th Ed., Washington, D.C., p. 9-18.

3. a.   *Enterobacter aerogenes*

    Refer to American Public Health Association; American Water Works Association; and Water Environment Federation (1992) *Standard Methods for the Examination of Water and Wastewater*. 18th Ed., Washington, D.C., pp. 9-5 and 9-6.

4. c.   170 °C for at least 1 hour

    Refer to American Public Health Association; American Water Works Association; and Water Environment Federation (1992) *Standard Methods for the Examination of Water and Wastewater*. 18th Ed., Washington, D.C., p. 9-16.

5. c. 44.5 °C ± 0.2 °C

Refer to American Public Health Association; American Water Works Association; and Water Environment Federation (1992) *Standard Methods for the Examination of Water and Wastewater.* 18th Ed., Washington, D.C., p. 9-60.

6. c. 20 to 60 CFU/100 mL

Refer to American Public Health Association; American Water Works Association; and Water Environment Federation (1992) *Standard Methods for the Examination of Water and Wastewater.* 18th Ed., Washington, D.C., p. 9-61.

7. b. 121 °C

Refer to American Public Health Association; American Water Works Association; and Water Environment Federation (1992) *Standard Methods for the Examination of Water and Wastewater.* 18th Ed., Washington, D.C., p. 9-14.

8. c. Gram-positive

Refer to American Public Health Association; American Water Works Association; and Water Environment Federation (1992) *Standard Methods for the Examination of Water and Wastewater.* 18th Ed., Washington, D.C., p. 9-45.

9. b. Should be a pathogen

Refer to American Public Health Association; American Water Works Association; and Water Environment Federation (1992) *Standard Methods for the Examination of Water and Wastewater.* 18th Ed., Washington, D.C., p. 9-1.

10. c. Calculated

Refer to American Public Health Association; American Water Works Association; and Water Environment Federation (1992) *Standard Methods for the Examination of Water and Wastewater.* 18th Ed., Washington, D.C., p. 9-45.

11. b. Lauryl tryptose broth, incubator, and autoclave

Lauryl tryptose broth is used for the presumptive phase of the coliform test. An incubator set at 35 °C is used for the presumptive, confirmed, and completed phases. The autoclave at 121 °C for 15 minutes is used for sterilization during the completed phase.

Refer to American Public Health Association; American Water Works Association; and Water Environment Federation (1992) *Standard Methods for the Examination of Water and Wastewater.* 18th Ed., Washington, D.C., p. 9-46.

12. b. Chemical and physical tests do not assess the effect on aquatic biota

Refer to American Public Health Association; American Water Works Association; and Water Environment Federation (1992) *Standard Methods for the Examination of Water and Wastewater.* 18th Ed., Washington, D.C., p. 8-1; and Smith, R.-K. (1999) *Lectures on Wastewater Analysis and Interpretation.* Genium Publishing, Schenectady, N.Y., p. 345.

13. b. Screening test

Refer to American Public Health Association; American Water Works Association; and Water Environment Federation (1992) *Standard Methods for the Examination of Water and Wastewater.* 18th Ed., Washington, D.C., p. 8-2; and Smith, R.-K. (1999) *Lectures on Wastewater Analysis and Interpretation.* Genium Publishing, Schenectady, N.Y., p. 347.

14. a. Acute test

Refer to American Public Health Association; American Water Works Association; and Water Environment Federation (1992) *Standard Methods for the Examination of Water and Wastewater.* 18th Ed., Washington, D.C., p. 8-2; and Smith, R.-K. (1999) *Lectures on Wastewater Analysis and Interpretation.* Genium Publishing, Schenectady, N.Y., p. 346.

15. b. Chronic test

Refer to American Public Health Association; American Water Works Association; and Water Environment Federation (1992) *Standard Methods for the Examination of Water and Wastewater.* 18th Ed., Washington, D.C., p. 8-2; and Smith, R.-K. (1999) *Lectures on Wastewater Analysis and Interpretation.* Genium Publishing, Schenectady, N.Y., p. 346.

16. a. Static test

Refer to American Public Health Association; American Water Works Association; and Water Environment Federation (1992*) Standard Methods for the Examination of Water and Wastewater.* 18th Ed., Washington, D.C., p. 8-4; and Smith, R.-K. (1999) *Lectures on Wastewater Analysis and Interpretation.* Genium Publishing, Schenectady, N.Y., p. 347.

17. a. LC

This is usually defined as the median (50%) lethal concentration (LC_{50}).

Refer to American Public Health Association; American Water Works Association; and Water Environment Federation (1992) *Standard Methods for the Examination of Water and Wastewater.* 18th Ed., Washington, D.C., p. 8-2; and

Smith, R.-K. (1999) *Lectures on Wastewater Analysis and Interpretation.* Genium Publishing, Schenectady, N.Y., p. 346.

18.   c.   IC

For examination, an IC_{25} could be the concentration estimated to cause a 25% reduction in the growth of larval fish relative to the control.

Refer to American Public Health Association; American Water Works Association; and Water Environment Federation (1992) *Standard Methods for the Examination of Water and Wastewater.* 18th Ed., Washington, D.C., p. 8-2; and Smith, R.-K. (1999) *Lectures on Wastewater Analysis and Interpretation.* Genium Publishing, Schenectady, N.Y., p. 346.

19.   a.   Dose

Refer to American Public Health Association; American Water Works Association; and Water Environment Federation (1992) *Standard Methods for the Examination of Water and Wastewater.* 18th Ed., Washington, D.C., p. 8-2; and Smith, R.-K. (1999) *Lectures on Wastewater Analysis and Interpretation.* Genium Publishing, Schenectady, N.Y., p. 345.

20.   c.   Control

Refer to American Public Health Association; American Water Works Association; and Water Environment Federation (1992) *Standard Methods for the Examination of Water and Wastewater.* 18th Ed., Washington, D.C., p. 8-2; and Smith, R.-K. (1999) *Lectures on Wastewater Analysis and Interpretation.* Genium Publishing, Schenectady, N.Y., p. 351.

21. b. Renewal test

Refer to American Public Health Association; American Water Works Association; and Water Environment Federation (1992) *Standard Methods for the Examination of Water and Wastewater.* 18th Ed., Washington, D.C. p. 8-3; and Smith, R.-K. (1999) *Lectures on Wastewater Analysis and Interpretation.* Genium Publishing, Schenectady, N.Y., p. 347.

22. d. LOEC

Refer to American Public Health Association; American Water Works Association; and Water Environment Federation (1992) *Standard Methods for the Examination of Water and Wastewater.* 18th Ed., Washington, D.C., p. 8-2; and Smith, R.-K. (1999) *Lectures on Wastewater Analysis and Interpretation.* Genium Publishing, Schenectady, N.Y., p. 347.

23. a. Short-term toxicity
The effluent may have a longer-term toxicity that will not be determined by this test.

Refer to American Public Health Association; American Water Works Association; and Water Environment Federation (1992*) Standard Methods for the Examination of Water and Wastewater.* 18th Ed., Washington, D.C., p. 8-4; and Smith, R.-K. (1999) *Lectures on Wastewater Analysis and Interpretation.* Genium Publishing, Schenectady, N.Y., p. 353.

24. d. Different species have various levels of sensitivity to different toxicants
A relative sensitivity test can be performed to determine which species is appropriate for an individual toxicity test.

Refer to American Public Health Association; American Water Works Association; and Water Environment Federation (1992) *Standard Methods for the Examination of Water and Wastewater*. 18th Ed., Washington, D.C., p. 8-5; and Smith, R.-K. (1999) *Lectures on Wastewater Analysis and Interpretation*. Genium Publishing, Schenectady, N.Y., p. 345.

25. b. Fathead minnows

This species is the fathead minnow, which can be used in both chronic and acute tests. The endpoint is death, and growth can also be measured in the chronic test.

Refer to American Public Health Association; American Water Works Association; and Water Environment Federation (1992) *Standard Methods for the Examination of Water and Wastewater*. 18th Ed., Washington, D.C., p. 8-80; and Smith, R.-K. (1999) *Lectures on Wastewater Analysis and Interpretation*. Genium Publishing, Schenectady, N.Y., p. 349.

26. c. 10%

Refer to American Public Health Association; American Water Works Association; and Water Environment Federation (1992) *Standard Methods for the Examination of Water and Wastewater*. 18th Ed., Washington, D.C., p. 9-18.

27. a. 1.25 mL KH_2PO_4 buffer and 5 mL $MgCl_2$ solution

Refer to American Public Health Association; American Water Works Association; and Water Environment Federation (1992) *Standard Methods for the Examination of Water and Wastewater*. 18th Ed., Washington, D.C., p. 9-18.

28. c. Place 1 mL of sample in 99 mL of blank and run 10 mL of inoculated blank across a filter

Refer to American Public Health Association; American Water Works Association; and Water Environment Federation (1992) *Standard Methods for the Examination of Water and Wastewater.* 18th Ed., Washington, D.C., p. 9-18.

29. c. Separate sterile pipets for each transfer

Refer to American Public Health Association; American Water Works Association; and Water Environment Federation (1992) *Standard Methods for the Examination of Water and Wastewater.* 18th Ed., Washington, D.C., p. 9-35.

30. d. 7

Refer to U.S. Code of Federal Regulations, *Standards for the Use and Disposal of Sewage Sludge.* 40 CFR, Parts 257, 403, and 503, Washington, D.C., 503.32(b) 2.

31. a. CFU/g

Refer to U.S. Code of Federal Regulations, *Standards for the Use and Disposal of Sewage Sludge.* 40 CFR, Parts 257, 403, and 503, Washington, D.C., 503.32(b) 2.

32. c. 100X

Refer to American Public Health Association; American Water Works Association; and Water Environment Federation (1992) *Standard Methods for the Examination of Water and Wastewater.* 18th Ed., Washington, D.C., p. 9-39.

33. a. Blue

Refer to American Public Health Association; American Water Works Association; and Water Environment Federation (1992) *Standard Methods for the Examination of Water and Wastewater*. 18th Ed., Washington, D.C., p. 9-61.

34. c. Gold–green (sheen)

Refer to American Public Health Association; American Water Works Association; and Water Environment Federation (1992) *Standard Methods for the Examination of Water and Wastewater*. 18th Ed., Washington, D.C., p. 9-56.

35. d. 20 to 80

Refer to American Public Health Association; American Water Works Association; and Water Environment Federation (1992) *Standard Methods for the Examination of Water and Wastewater*. 18th Ed., Washington, D.C., p. 9-57.

36. c. 20 to 60

Refer to American Public Health Association; American Water Works Association; and Water Environment Federation (1992) *Standard Methods for the Examination of Water and Wastewater*. 18th Ed., Washington, D.C., p. 9-61.

37. b. Gas bubbles

Refer to American Public Health Association; American Water Works Association; and Water Environment Federation (1992) *Standard Methods for the Examination of Water and Wastewater*. 18th Ed., Washington, D.C., p. 9-47.

38. c. Sterilize

Refer to American Public Health Association; American Water Works Association; and Water Environment Federation (1992) *Standard Methods for the Examination of Water and Wastewater.* 18th Ed., Washington, D.C., p. 1-39.

39. c. Pathogens

Refer to Water Environment Federation (1996) *Operation of Municipal Wastewater Treatment Plants.* 5th Ed., Manual of Practice No. 11, Alexandria, Va., p. 481.

40. b. Remove any residual chlorine

Refer to American Public Health Association; American Water Works Association; and Water Environment Federation (1992) *Standard Methods for the Examination of Water and Wastewater.* 18th Ed., Washington, D.C., p. 9-18.

41. c. 44.5 °C

Refer to American Public Health Association; American Water Works Association; and Water Environment Federation (1992) *Standard Methods for the Examination of Water and Wastewater.* 18th Ed., Washington, D.C., p. 9-61.

PROCESS CONTROL

Questions

1. SVI compares which two parameters?

    a. Phosphorus level to nitrogen level
    b. Settling of MLSS to concentration of MLSS
    c. Semivolatile index to MLSS
    d. Return sludge settling to wasting rate

2. The mixed liquor settleability test shows settling characteristics under controlled conditions. Why is a settleometer preferred over a 1000-mL graduated cylinder for this test?

    a. Sidewall effects of a narrow cylinder interfere with settling
    b. Microorganisms settle better when they think they are in a clarifier
    c. There is more room for filaments to spread in a wide beaker
    d. The settleometer is easier for the analyst to handle

3. During the conduct of the settleable solids test, in some samples, a separation of settleable and floating material may occur. In such instances,

    a. The weight (or volume) of floating and settled material are added together and reported
    b. The weight (or volume) of floating material is subtracted from that of settled material and reported
    c. Only the weight (or volume) of floating material is reported
    d. Only the weight (or volume) of settled material is reported

4. When conducting a settleable solids test, if settled material develops pockets of liquid between large settled particles,

    a. Consider all materials above the lowest settled portion as floating material and disregard
    b. Estimate the volume of the liquid pockets and subtract from the total volume of settled solids
    c. Estimate the volume of the liquid pockets and add to the total volume of settled solids
    d. Ignore the liquid pockets and report settled solids as usual

5. Settleable solids may be determined

    a. On a volumetric basis and reported as milligrams per liter
    b. On a volumetric basis and reported as milliliters per kilogram
    c. On a gravimetric basis and reported as milligrams per liter
    d. On a gravimetric basis and reported as milligrams per kilogram

6. The procedure for determining settleable solids by the volumetric method may be summarized as

    a. Recording the level of settled solids in milligrams per liter after a 45-minute settling period
    b. Recording the level of settled solids in milligrams per liter after a 30-minute settling period
    c. Recording the levels of settled solids in milliliters per liter after a 30 minutes of settling followed by gentle agitation (near the sides of the cone) and an additional 15-minute settling period
    d. Recording the level of settled solids in milliliters per liter after 45 minutes of settling followed by gentle agitation (near the sides of the cone) and an additional 15-minute settling period

7. Microscopic examination of activated sludge is

    a. Not usually important
    b. A good way to identify toxic substances
    c. Helpful in identifying pathogens
    d. Helpful in indicating adequate or poor treatment

8. MLSS is important for

    a. Preliminary treatment process control
    b. Calculating BOD
    c. Calculating MCRT
    d. Controlling trickling filter effluent

9. What can you determine by centrifuging sludge samples?

    a. Concentration by volume
    b. Concentration by weight
    c. Density
    d. Specific gravity

10. Which indicates a good quality domestic activated sludge?

    a. Black color and septic odor
    b. Brown color and musty odor
    c. Brown color and lots of dark brown foam
    d. Settleability of 900 mL in a 1-L cylinder during a 30-minute period

11. Calculate SVI from the following laboratory data.

    The sludge level in a 1-L graduated cylinder after 30 minutes of settling is 195 mL and SS in the mixed liquor is 2300 mg/L.

    a. 85
    b. 100
    c. 117
    d. 147

12. SVI is

    a. The weight in milligrams of 1 mL of MLSS after 30 minutes of settling
    b. The volume of sludge blanket divided by the daily volume of sludge pumped from the thickener
    c. The volume in milliliters occupied by 1 g of MLSS after 30 minutes of settling
    d. The weight in kilograms of 1 mL of MLSS after 60 minutes of settling

13. During a specific oxygen uptake rate analysis, DO readings are taken at time intervals no less than

    a. 30 seconds
    b. 1 minute
    c. 15 minutes
    d. 1 hour

14. Turbidity in wastewater is caused by

    a. Color
    b. Colloidal particles

c. Dissolved calcium
d. Hardness

15. Jar tests are used to determine the optimum dosage of

   a. Chlorine
   b. SO_2
   c. Recycled effluent
   d. $Al_2(SO_4)_3$

Answers

1. b. Settling of MLSS to concentration of MLSS

    Refer to American Public Health Association; American Water Works Association; and Water Environment Federation (1992) *Standard Methods for the Examination of Water and Wastewater*. 18th Ed., Washington, D.C., p. 2-66.

2. a. Sidewall effects of a narrow cylinder interfere with settling
   A wider area better simulates a secondary clarifier.

    Refer to Water Environment Federation (1996) *Operation of Municipal Wastewater Treatment Plants*. 5th Ed., Manual of Practice No. 11, Alexandria, Va., p. 648.

3. d. Only the weight (or volume) of settled material is reported

    Refer to American Public Health Association; American Water Works Association; and Water Environment Federation (1992) *Standard Methods for the Examination of Water and Wastewater*. 18th Ed., Washington, D.C., p. 2-57.

4. b. Estimate the volume of the liquid pockets and subtract from the total volume of settled solids

    Refer to American Public Health Association; American Water Works Association; and Water Environment Federation (1992) *Standard Methods for the Examination of Water and Wastewater*. 18th Ed., Washington, D.C., p. 2-57.

5. c. On a gravimetric basis and reported as milligrams per liter

    Refer to American Public Health Association; American Water Works Association; and Water Environment Federation (1992) *Standard Methods for the Examination of Water and Wastewater.* 18th Ed., Washington, D.C., p. 2-57.

6. d. Recording the level of settled solids in milliliters per liter after 45 minutes of settling followed by gentle agitation (near the sides of the cone) and an additional 15-minute settling period

    Refer to American Public Health Association; American Water Works Association; and Water Environment Federation (1992) *Standard Methods for the Examination of Water and Wastewater.* 18th Ed., Washington, D.C., p. 2-57.

7. d. Helpful in indicating adequate or poor treatment
    It is not necessary to be a skilled microbiologist or to be able to identify or count individual species. Rather, the operator or technician needs only to recognize primary groups of organisms such as protozoa, rotifers, filamentous bacteria, and nematodes.

    Refer to Water Environment Federation (1996) *Operation of Municipal Wastewater Treatment Plants.* 5th Ed., Manual of Practice No. 11, Alexandria, Va., p. 632; and Smith, R.-K. (1999) *Lectures on Wastewater Analysis and Interpretation.* Genium Publishing, Schenectady, N.Y., p. 341.

8. c. Calculating MCRT

    Refer to Water Environment Federation (1996) *Operation of Municipal Wastewater Treatment Plants.* 5th Ed., Manual of Practice No. 11, Alexandria, Va., Chap. 20.

9. a. Concentration by volume

Refer to California State University (1992) *Operation of Wastewater Treatment Plants*, 4th Ed., Sacramento, Calif., p. 486.

10. b. Brown color and musty odor

Refer to California State University (1992) *Operation of Wastewater Treatment Plants.* 4th Ed., Sacramento, Calif., p. 271.

11. a. 85

$$\text{SVI, g/mL} = \frac{\text{(Settleable Sludge Volume, mL/L)} \times \text{(1000 mL/L)}}{\text{Suspended Solids, mg/L}}$$

Refer to American Public Health Association; American Water Works Association; and Water Environment Federation (1992) *Standard Methods for the Examination of Water and Wastewater.* 18th Ed., Washington, D.C., p. 2-66.

12. c. The volume in milliliters occupied by 1 g of MLSS after 30 minutes of settling

Refer to American Public Health Association; American Water Works Association; and Water Environment Federation (1992) *Standard Methods for the Examination of Water and Wastewater.* 18th Ed., Washington, D.C., p. 2-66.

13. b. 1 minute

Refer to American Public Health Association; American Water Works Association; and Water Environment Federation (1992) *Standard Methods for the Examination of Water and Wastewater.* 18th Ed., Washington, D.C., p. 2-64.

14. b. Colloidal particles

Refer to American Public Health Association; American Water Works Association; and Water Environment Federation (1992) *Standard Methods for the Examination of Water and Wastewater*. 18th Ed., Washington, D.C., p. 2-8.

15. d. $Al_2(SO_4)_3$

Refer to Water Environment Federation (1996) *Operation of Municipal Wastewater Treatment Plants*. 5th Ed., Manual of Practice No. 11, Alexandria, Va., p. 478.

RESIDUE

Questions

1. The practical range of determination for dissolved solids is

    a. 10 to 20 000 mg/L
    b. 10 to 200 000 mg/L
    c. 1000 to 20 000 mg/L
    d. 1000 to 200 000 mg/L

2. Filterable residue is another name for

    a. TSS
    b. Dissolved solids
    c. Total solids
    d. Fixed solids

3. A well-mixed sample is filtered through a standard glass fiber filter. The filtrate is evaporated and dried to a constant weight at 180 °C. This summarizes the U.S. EPA method for determining

    a. Filterable residue
    b. Nonfilterable residue
    c. Volatile residue
    d. Total filterable and nonfilterable residue

4. Theoretical dissolved solids concentration is calculated by

    a. Taking the difference between the total solids concentration and the SS concentration on the same sample

b. Taking the difference between the total solids concentration and the volatile solids concentration on the same sample

c. Taking the difference between the total volatile solids concentration and the total filterable volatile solids concentration on the same sample

d. Taking the difference between the filterable solids concentration and nonfilterable solids concentration on the same sample

5. The practical range of determination for U.S. EPA Method 160.2 (Residue, Nonfilterable) is

   a. 1 to 20 000 mg/L
   b. 4 to 20 000 mg/L
   c. 100 to 20 000 mg/L
   d. 1000 to 20 000 mg/L

6. Nonfilterable residue is defined as those solids

   a. Passed through a glass fiber filter and dried to a constant weight at 103 to 105 °C
   b. Retained by a glass fiber filter and dried to a constant weight at 103 to 105 °C
   c. Retained by a glass fiber filter and dried to a constant weight at 180 °C
   d. Retained by a glass fiber filter and burned away to a constant weight at 550 °C

7. Which of the following is NOT part of the preparation of glass fiber filters?

   a. Placing filters in a Gooch crucible and applying suction
   b. Rinsing once with 10 mL distilled water and applying suction until dry

c. Rinsing with three successive 20-mL volumes of distilled water and continuing suction until all visible traces of water are removed

d. Handling rinsed, dried, and weighed filters/crucibles with forceps or tongs only

8. During the performance of a TSS test, filters with residue are dried for at least 1 hour, cooled in a desiccator, and weighed. This cycle is repeated until a constant weight is obtained. Constant weight is defined as

   a. A weight gain of less than 1 mg
   b. A weight loss of less than 1 mg
   c. A weight difference (±) of less than 0.5 mg
   d. A weight difference (±) of less than 5.0 mg

9. Reduce TSS sample volume if the filtration time exceeds

   a. 5 to 10 minutes
   b. 10 to 15 minutes
   c. 15 to 20 minutes
   d. 20 to 25 minutes

10. Total solids are

    a. The solids remaining on the filter after filtration
    b. The sum of TSS and dissolved solids
    c. Filterable solids
    d. Nonfilterable solids

11. Total solids are determined based on evaporation

    a. At 103 °C

b. At 103 to 105 °C

c. In an oven at 98 °C, initially, then 1 hour at 103 to 105 °C

d. At 98 °C

12. Prepare clean evaporating dishes for total residue by

    a. Heating in a drying oven at 103 to 105 °C for 1 hour, then cooling and storing in a desiccator
    b. Heating in a drying oven at 103 to 105 °C for 1 hour, then cooling and storing in a clean area
    c. Heating in a muffle furnace at 550 °C for 1 hour, then cooling and storing in a desiccator
    d. Heating in a drying oven at 180 °C for 1 hour, then cooling and storing in a desiccator

13. Total volatile solids represent the weight of material lost

    a. After burning the total solids component at 550 °C
    b. After drying the total solids at 103 to 105 °C
    c. After drying the total solids at 180 °C
    d. After burning the filter and solids at 550 °C

14. TSS is another name for

    a. Filterable residue
    b. Nonfilterable residue
    c. Total residue
    d. Total and filterable residue

15. In a TSS analysis, constant weight is achieved when

    a. Weight change is less than 2% of the previous weight
    b. Weight change is less than 4% of the previous weight
    c. Weight change is less than 5% of the previous weight
    d. Weight change is less than 10% of the previous weight

16. What is the proper temperature of a drying oven used in the determination of SS?

    a. 104 °F ± 1 °F
    b. 104 °C ± 1 °C
    c. 180 °F ± 1 °F
    d. 180 °C ± 1 °C

17. At what temperature is a muffle oven set to determine volatile solids?

    a. 180 °F ± 50 °F
    b. 180 °C ± 50 °C
    c. 550 °F ± 50 °F
    d. 550 °C ± 50 °C

18. TSS refers to

    a. Total solids minus settleable solids
    b. Material retained after passing the sample through a standard glass fiber filter
    c. Material remaining in a sample container after drying at 103 to 105 °C
    d. Solids remaining in a sample after centrifuging

19. Fixed solids are defined as solids that

    a. Pass though a glass fiber filter
    b. Are lost upon ignition at 550 °C ± 50 °C
    c. Remain fixed on a glass fiber filter after drying at 105 °C
    d. Remain after ignition at 550 °C ± 50 °C

20. What is the correct temperature for a drying oven used in the determination of suspended and total solids?

    a. 100 °F ± 1 °F
    b. 104 °C ± 1 °C
    c. 180 °F ± 1 °F
    d. 212 °C ± 1 °C

21. A piece of equipment used in the total residue determination is the

    a. Friedrichs condenser
    b. Gooch crucible
    c. Evaporating dish
    d. Separatory funnel

22. Upon weighing a total filterable residue sample after evaporative drying and cooling, a technician finds that the ending mass is less than the mass of the dish itself. Which of the following could account for this?

    a. The tare of the balance was not checked before weighing the sample
    b. The sample was contaminated with phlogiston
    c. The dish was sufficiently cooled before weighing the sample
    d. The sample had been heated before filtering

23. To what does SS in wastewater refer?

    a. All solids in wastewater
    b. Solids retained on a filter
    c. Fixed solids
    d. Volatile solids

24. How is the theoretical amount of dissolved solids in raw wastewater determined?

    a. Adding settleable solids to the volatile solids
    b. Running the total solids test
    c. Subtracting settleable solids from the total solids
    d. Subtracting SS from the total solids

25. According to U.S. EPA Method 160.3 (Residue, Total), a sample volume for total solids analysis should contain a residue of at least

    a. 25 mg
    b. 50 mg
    c. 100 mg
    d. 200 mg

26. According to U.S. EPA Method 160.2 (Residue, Nonfilterable), sufficient sample volume for TSS analysis will result in a minimum of how much residue on a 4.7-cm filter?

    a. 0.1 mg
    b. 1.0 mg
    a. 10 mg
    b. 20 mg

27. Ash content of a sample is the same as

   a. Fixed residue
   b. Nonfilterable residue
   c. Volatile solids
   d. Organic solids

Answers

1. a. 10 to 20 000 mg/L

    Refer to U.S. Environmental Protection Agency (1983) *Methods for Chemical Analysis of Water and Wastewater.* EPA-600/4-79-020 (revised March 1983), Environ. Monit. Support Lab., Cincinnati, Ohio, p. 160.1-1.

2. b. Dissolved solids

    Refer to U.S. Environmental Protection Agency (1983) *Methods for Chemical Analysis of Water and Wastewater.* EPA-600/4-79-020 (revised March 1983), Environ. Monit. Support Lab., Cincinnati, Ohio, p. 160.1-1.

3. a. Filterable residue

    Refer to U.S. Environmental Protection Agency (1983) *Methods for Chemical Analysis of Water and Wastewater.* EPA-600/4-79-020 (revised March 1983), Environ. Monit. Support Lab., Cincinnati, Ohio, p. 160.1-1.

4. a. Taking the difference between the total solids concentration and the SS concentration on the same sample

    Refer to American Public Health Association; American Water Works Association; and Water Environment Federation (1992) *Standard Methods for the Examination of Water and Wastewater.* 18th Ed., Washington, D.C., p. 2-55.

5. b. 4 to 20 000 mg/L

    Refer to U.S. Environmental Protection Agency (1983) *Methods for Chemical Analysis of Water and Wastewater.* EPA-600/4-79-020 (revised March 1983), Environ. Monit. Support Lab., Cincinnati, Ohio, p. 160.2-1.

6. b. Retained by a glass fiber filter and dried to a constant weight at 103 to 105 °C

    Refer to U.S. Environmental Protection Agency (1983) *Methods for Chemical Analysis of Water and Wastewater.* EPA-600/4-79-020 (revised March 1983), Environ. Monit. Support Lab., Cincinnati, Ohio, p. 160.2-1.

7. b. Rinsing once with 10 mL distilled water and applying suction until dry

    Refer to U.S. Environmental Protection Agency (1983) *Methods for Chemical Analysis of Water and Wastewater.* EPA-600/4-79-020 (revised March 1983), Environ. Monit. Support Lab., Cincinnati, Ohio, p. 160.2-2.

8. c. A weight difference (±) of less than 0.5 mg

    Refer to U.S. Environmental Protection Agency (1983) *Methods for Chemical Analysis of Water and Wastewater.* EPA-600/4-79-020 (revised March 1983), Environ. Monit. Support Lab., Cincinnati, Ohio, p. 160.2-3.

9. a. 5 to 10 minutes

    Refer to U.S. Environmental Protection Agency (1983) *Methods for Chemical Analysis of Water and Wastewater.* EPA-600/4-79-020 (revised March 1983), Environ. Monit. Support Lab., Cincinnati, Ohio, p. 160.2-2.

10. b. The sum of TSS and dissolved solids

Refer to Water Environment Federation (1996) *Operation of Municipal Wastewater Treatment Plants.* 5th Ed., Manual of Practice No. 11, Alexandria, Va., p. 483.

11. c. In an oven at 98 °C initially, then 1 hour at 103 to 105 °C

Refer to U.S. Environmental Protection Agency (1983) *Methods for Chemical Analysis of Water and Wastewater.* EPA-600/4-79-020 (revised March 1983), Environ. Monit. Support Lab., Cincinnati, Ohio, p. 160.3-1.

12. a. Heating in a drying oven at 103 to 105 °C for 1 hour, then cooling and storing in a desiccator

Refer to U.S. Environmental Protection Agency (1983) *Methods for Chemical Analysis of Water and Wastewater.* EPA-600/4-79-020 (revised March 1983), Environ. Monit. Support Lab., Cincinnati, Ohio, p. 160.3-1.

13. a. After burning the total solids component at 550 °C

Refer to U.S. Environmental Protection Agency (1983) *Methods for Chemical Analysis of Water and Wastewater.* EPA-600/4-79-020 (revised March 1983), Environ. Monit. Support Lab., Cincinnati, Ohio, p. 160.4-1.

14. b. Nonfilterable residue

Refer to U.S. Environmental Protection Agency (1983) *Methods for Chemical Analysis of Water and Wastewater.* EPA-600/4-79-020 (revised March 1983), Environ. Monit. Support Lab., Cincinnati, Ohio, p. 160.2-1.

15. b. Weight change is less than 4% of the previous weight

Refer to American Public Health Association; American Water Works Association; and Water Environment Federation (1992) *Standard Methods for the Examination of Water and Wastewater.* 18th Ed., Washington, D.C., p. 2-56.

16. b. 104 °C ± 1 °C

Refer to American Public Health Association; American Water Works Association; and Water Environment Federation (1992) *Standard Methods for the Examination of Water and Wastewater.* 18th Ed., Washington, D.C., p. 2-56.

17. d. 550 °C ± 50 °C

Refer to American Public Health Association; American Water Works Association; and Water Environment Federation (1992) *Standard Methods for the Examination of Water and Wastewater.* 18th Ed., Washington, D.C., p. 2-57.

18. b. Material retained after passing the sample through a standard glass fiber filter

Refer to American Public Health Association; American Water Works Association; and Water Environment Federation (1992) *Standard Methods for the Examination of Water and Wastewater.* 18th Ed., Washington, D.C., p. 2-53.

19. d. Remain after ignition at 550 °C ± 50 °C

Refer to American Public Health Association; American Water Works Association; and Water Environment Federation (1992) *Standard Methods for the Examination of Water and Wastewater.* 18th Ed., Washington, D.C., p. 2-53.

20. b. 104 °C ± 1 °C

Refer to American Public Health Association; American Water Works Association; and Water Environment Federation (1992) *Standard Methods for the Examination of Water and Wastewater.* 18th Ed., Washington, D.C., p. 2-54.

21. c. Evaporating dish

Refer to American Public Health Association; American Water Works Association; and Water Environment Federation (1992) *Standard Methods for the Examination of Water and Wastewater.* 18th Ed., Washington, D.C., p. 2-54.

22. a. The tare of the balance was not checked before weighing the sample

Refer to American Public Health Association; American Water Works Association; and Water Environment Federation (1992) *Standard Methods for the Examination of Water and Wastewater.* 18th Ed., Washington, D.C., p. 2-54.

23. b. Solids retained on a filter

Refer to American Public Health Association; American Water Works Association; and Water Environment Federation (1992) *Standard Methods for the Examination of Water and Wastewater.* 18th Ed., Washington, D.C., p. 2-53.

24. d. Subtracting the SS from the total solids

Refer *to* American Public Health Association; American Water Works Association; and Water Environment Federation (1992) *Standard Methods for the Examination of Water and Wastewater.* 18th Ed., Washington, D.C., p. 2-53.

25. a. 25 mg

Refer to U.S. Environmental Protection Agency (1983) *Methods for Chemical Analysis of Water and Wastewater.* EPA-600/4-79-020 (revised March 1983), Environ. Monit. Support Lab., Cincinnati, Ohio, p. 160.3-1.

26. b. 1.0 mg

Refer to U.S. Environmental Protection Agency (1983) *Methods for Chemical Analysis of Water and Wastewater.* EPA-600/4-79-020 (revised March 1983), Environ. Monit. Support Lab., Cincinnati, Ohio, p. 160.2-2.

27. a. Fixed residue

Refer to American Public Health Association; American Water Works Association; and Water Environment Federation (1992) *Standard Methods for the Examination of Water and Wastewater.* 18th Ed., Washington, D.C., p. 2-53.

GENERAL

Questions

1. When using volumetric glassware, the reading should be taken

    a. At the top of the meniscus
    b. At the bottom of the meniscus
    c. At the middle of the meniscus
    d. The meniscus has no bearing on the reading

2. Beakers should NOT be used for

    a. Mixing
    b. Weighing chemicals
    c. Heating solutions
    d. Measuring exact volumes

3. What can cause sampling errors?

    a. Proper sampling
    b. Uncalibrated equipment
    c. Good preservation
    d. Thorough mixing during compositing and testing

4. Laboratory and field thermometers or other temperature-measuring devices should be calibrated at least once per year against

    a. Another manufacturer's thermometer
    b. Another laboratory's thermometer
    c. A NIST-certified thermometer
    d. A HTC-certified thermometer

5. Temperature results are reported to the nearest

    a. Whole degree in Fahrenheit or Celsius, as required
    b. Whole degree in Fahrenheit
    c. Whole degree in Celsius
    d. 0.1 or 1.0 °C, depending on the application

6. Preservation and hold time recommendations for samples collected for hardness analysis are

    a. Acidified with HNO_3 to pH < 2.0 with a maximum hold time of 14 days
    b. Acidified with HNO_3 to pH < 2.0 with a maximum hold time of 28 days
    c. Acidified with HNO_3 to pH < 2.0 with a maximum hold time of 6 months
    d. Cooled to 4 °C with a maximum hold time of 6 months

7. Analytical results for turbidity are expressed as

    a. milligrams per liter
    b. standard units
    c. millivolts
    d. nephelometric turbidity units

8. Turbidity is defined as an expression of

    a. The optical property that causes light to be scattered and absorbed rather than transmitted in straight lines through the sample
    b. A correlation between TSS concentration and color
    c. A correlation between dissolved solids concentration plus TSS concentration and color
    d. The optical property that causes light to stratify in a sample

9. Which of the following could be a possible interference in turbidity readings?

    a. Dirty, scratched glassware
    b. A high COD concentration
    c. High cyanide or phenol concentrations
    d. Filamentous bacteria

10. An insoluble compound formed by the combination of two or more soluble materials is called a

    a. Coagulant
    b. Precipitate
    c. Filter aid
    d. Sequestering agent

11. At what temperature is water at its greatest density?

    a. 4 °C
    b. 15 °C
    c. 39 °C
    d. 48 °F

12. A normal solution is defined as

    a. A solution containing 1 g molecular weight to the volume of distilled solute per liter of solution
    b. The volume of concentrated solute diluted to the volume of distilled water required for dilution
    c. A solution containing 1 g equivalent weight of solute per liter of solution
    d. A standard substance that is accompanied by a certificate of analysis

13. The density of water at 4 °C is

    a. 1.000 g/cm³
    b. 10.00 g/cm³
    c. 100.0 g/cm³
    d. 1000.0 g/cm³

14. Filtration is a

    a. Chemical process
    b. Physical process
    c. Thermal process
    d. Biological process

15. A process in which the liquid is vaporized, recondensed, and collected is

    a. Digestion
    b. Distillation
    c. Dilution
    d. Deionization

16. At what temperature does water freeze?

    a. 0 °C
    b. 0 °F
    c. 32 °C
    d. 100 °F

17. An example of a base is

    a. NH_4OH

b. NH₄Cl
c. (NH₄)₂SO₄
d. AlCl₃

18. Which of the following best describes septic wastewater?

    a. Black with a rotten egg odor
    b. Brown with a musty odor
    c. Gray with a musty odor
    d. Green and greasy

19. What do you call untreated wastewater?

    a. Fresh
    b. Ground
    c. Raw
    d. Septic

20. The word *normal* associated with a reagent refers to the

    a. Gram equivalent weight of the reagent
    b. Molecular weight of the reagent
    c. Purity of the chemicals
    d. Reagent typically used

21. Buffer capacity is a measure of the ability to

    a. Hold settleable solids
    b. Resist change in pH
    c. Saturate DO
    d. Treat fluctuating COD load

22. Formazin standards are used to calibrate a

   a. Conductivity meter
   b. pH meter
   c. Oxygen meter
   d. Turbidity meter

23. A concentration of 1%, when the specific gravity is 1.00, is equal to

   a. 1 mg/L
   b. 10 mg/L
   c. 1000 mg/L
   d. 10 000 mg/L

24. Sulfur is found commonly in one of four valence states. In which valence state is sulfur present in the sulfite ion?

   a. −2
   b. 0
   c. +4
   d. +6

25. If you prepare a $K_2Cr_2O_7$ solution of 1 M and another of 1 N concentration, which of the following describes these concentrations?

   a. They are the same
   b. The molar solution is three times less concentrated than the normal solution
   c. The normal solution is six times less concentrated than the molar solution
   d. The normal solution is three times less concentrated than the molar solution

26. An example of a test used to measure a nutrient level is

    a. COD
    b. Total phosphorus
    c. Fecal coliform
    d. Total solids

27. An example of a test used to detect toxic organic pollutants is

    a. COD
    b. Total phosphorus by single-reagent molybdate colorimetry
    c. Fecal coliform by membrane filtration
    d. Gas chromatography/mass spectrometry

28. Which of the following expressions represents "micrograms per liter"?

    a. mg/L
    b. ng/L
    c. mL/L
    d. µg/L

29. Hardness results are expressed as

    a. milligrams per liter
    b. milligrams per liter as $CaCO_3$
    c. micrograms per liter
    d. micrograms per liter as $CaCO_3$

30. How long after buffer addition should the titration of hardness samples be completed?

    a. 3 minutes
    b. 5 minutes
    c. 8 minutes
    d. 10 minutes

31. The sample volume used for hardness titration should be adjusted so that the volume of EDTA titrant used is less than

    a. 10 mL
    b. 15 mL
    c. 25 mL
    d. 50 mL

32. The recommended preservation steps and holding times for samples collected for color analysis are

    a. Collect and acidify with HNO_3 to pH < 2.0 with a maximum hold time of 48 hours
    b. Collect in glass or plastic and cool to 4 °C with a maximum hold time of 48 hours
    c. Collect in glass only and cool to 4 °C with a maximum hold time of 2 hours
    d. Collect in glass only and cool to 4 °C with a maximum hold time of 24 hours

33. Which of these buffers will provide control at a pH of 7?

    a. NH_3/NH_4Cl
    b. Tris buffer

c. Na₂CO₃/NaHCO₃
d. KH₂PO₄/Na₂HPO₄

34. A "buffer" is a substance

    a. Added to a solution to change its pH
    b. Added to a solution to facilitate the determination of its pH
    c. That tends to inhibit changes in pH when an acid or base is added
    d. Added to culture media after titration to prevent pH fluctuation

35. The number of moles per liter of a solution is equivalent to

    a. Normality
    b. Concentration
    c. Formula weight
    d. Molarity

36. Which one of the following is an element?

    a. Lime
    b. Alum
    c. Hydrogen
    d. Salt

37. Which of the following is NOT a conventional pollutant?

    a. BOD
    b. TSS
    c. Fecal coliform
    d. Ammonia

Answers

1. b. At the bottom of the meniscus

   Refer to U.S. Environmental Protection Agency (1979) *Handbook for Analytical Quality Control in Water and Wastewater Laboratories.* EPA-600/4-79-019, Environ. Monit. Support Lab., Cincinnati, Ohio, p. 4-2.

2. d. Measuring exact volumes
   Beakers are not calibrated accurately enough to deliver accurate volumes.

   Refer to U.S. Environmental Protection Agency (1979) *Handbook for Analytical Quality Control in Water and Wastewater Laboratories.* EPA-600/4-79-019, Environ. Monit. Support Lab., Cincinnati, Ohio, p. 4-2.

3. b. Uncalibrated equipment

   Refer to American Public Health Association; American Water Works Association; and Water Environment Federation (1992) *Standard Methods for the Examination of Water and Wastewater.* 18th Ed., Washington, D.C., p. 1-18.

4. c. A NIST-certified thermometer

   Refer to American Public Health Association; American Water Works Association; and Water Environment Federation (1992) *Standard Methods for the Examination of Water and Wastewater.* 18th Ed., Washington, D.C., p. 2-59.

5. d. 0.1 or 1.0 °C, depending on the application

    Refer to American Public Health Association; American Water Works Association; and Water Environment Federation (1992) *Standard Methods for the Examination of Water and Wastewater.* 18th Ed., Washington, D.C., p. 2-59.

6. c. Acidified with HNO_3 to pH < 2.0 with a maximum hold time of 6 months

    Refer to U.S. Code of Federal Regulations (1999) *Guidelines Establishing Test Procedures for Analysis of Pollutants Under Clean Water Act.* 40 CFR, Part 136, Washington, D.C.

7. d. nephelometric turbidity units

    Refer to U.S. Environmental Protection Agency (1983) *Methods for Chemical Analysis of Water and Wastewater.* EPA-600/4-79-020 (revised March 1983), Environ. Monit. Support Lab., Cincinnati, Ohio, p. 180.1-1.

8. a. The optical property that causes light to be scattered and absorbed rather than transmitted in straight lines through the sample

    Refer to American Public Health Association; American Water Works Association; and Water Environment Federation (1992) *Standard Methods for the Examination of Water and Wastewater.* 18th Ed., Washington, D.C., p. 2-8.

9. a. Dirty, scratched glassware

    Refer to U.S. Environmental Protection Agency (1983) *Methods for Chemical Analysis of Water and Wastewater.* EPA-600/4-79-020 (revised March 1983), Environ. Monit. Support Lab., Cincinnati, Ohio, p. 180.1-1-2.

10. b. Precipitate

Water Environment Federation (1996) *Operation of Municipal Wastewater Treatment Plants.* 5th Ed., Manual of Practice No. 11, Alexandria, Va., p. 1322.

11. a. 4 °C

Refer to California State University (1992) *Operation of Wastewater Treatment Plants.* 4th Ed., Sacramento, Calif., p. 456.

12. c. A solution containing 1 g equivalent weight of solute per liter of solution

Refer to Smith, R.-K. (1995) *Water and Wastewater Laboratory Techniques.* Water Environ. Fed., Alexandria, Va., p. 199.

13. a. 1.000 g/cm^3

Refer to Smith, R.-K. (1995) *Water and Wastewater Laboratory Techniques.* Water Environ. Fed., Alexandria, Va., p. 174.

14. b. Physical process
Filtration is a physical process. Particles larger than the filter pores are retained on the surface; all other matter passes through.

Refer to Smith, R.-K. (1995) *Water and Wastewater Laboratory Techniques.* Water Environ. Fed., Alexandria, Va., p. 138.

15. b. Distillation

Refer to Smith, R.-K. (1995) *Water and Wastewater Laboratory Techniques.* Water Environ. Fed., Alexandria, Va., p. 127.

16. a.   0 °C

Refer to Smith, R.-K. (1995) *Water and Wastewater Laboratory Techniques.* Water Environ. Fed., Alexandria, Va., p. 92.

17. a. NH_4OH

Refer to Smith, R.-K. (1995) *Water and Wastewater Laboratory Techniques.* Water Environ. Fed., Alexandria, Va., p. 1.

18. a.   Black with a rotten egg odor

Refer to Water Environment Federation (1996) *Operation of Municipal Wastewater Treatment Plants.* 5th Ed., Manual of Practice No. 11, Alexandria, Va., p. 479.

19. c.   Raw

Refer to Water Environment Federation (1996) *Operation of Municipal Wastewater Treatment Plants.* 5th Ed., Manual of Practice No. 11, Alexandria, Va., p. 480.

20. a.   Gram equivalent weight of the reagent

Refer to Smith, R.-K. (1995) *Water and Wastewater Laboratory Techniques.* Water Environ. Fed., Alexandria, Va., p. 199.

21. b.   Resist change in pH

Refer to Smith, R.-K. (1999) *Handbook of Environmental Analysis.* 4th Ed., Genium Publishing, Schenectady, N.Y., p. 467.

22. d. Turbidity meter

Refer to American Public Health Association; American Water Works Association; and Water Environment Federation (1992) *Standard Methods for the Examination of Water and Wastewater.* 18th Ed., Washington, D.C., p. 2-10.

23. d. 10 000 mg/L

Refer to Smith, R.-K. (1999) *Handbook of Environmental Analysis.* 4th Ed., Genium Publishing, Schenectady, N.Y., p. 166.

24. c. +4

Sulfite (SO_3^{2-}) ion represents sulfur in the +4 valence state. Oxygen is always -2 and the total charge on the ion is -2, thus

$S + 3(-2) = -2$

$S = +4$

Refer to Smith, R.-K. (1995) *Water and Wastewater Laboratory Techniques.* Water Environ. Fed., Alexandria, Va., p. 202.

25. c. The normal solution is six times less concentrated than the molar solution Chromium in dichromate has an oxidation state of +6. Most reactions involving oxidation–reduction with chromium go from the +6 state to the +3 state, a three-electron change. There are two chromium atoms in dichromate $(Cr_2O_7)^{2-}$. Thus, each molecule of dichromate can provide six electrons. In most (not all) applications of dichromate then, normality will be six times less concentrated than molarity.

Refer to Smith, R.-K. (1995) *Water and Wastewater Laboratory Techniques.* Water Environ. Fed., Alexandria, Va., p. 197.

26. b. Total phosphorus

Phosphorus and nitrogen compounds are classified as nutrients.

Refer to Smith, R.-K. (1999) *Lectures on Wastewater Analysis and Interpretation.* Genium Publishing, Schenectady, N.Y., p. 187; and Smith, R.-K. (1999) *Handbook of Environmental Analysis.* 4th Ed., Genium Publishing, Schenectady, N.Y., p. 281.

27. d. Gas chromatography/mass spectrometry

Refer to Smith, R.-K. (1999) *Lectures on Wastewater Analysis and Interpretation.* Genium Publishing, Schenectady, N.Y., p. 75; and Smith, R.-K. (1999) *Handbook of Environmental Analysis.* 4th Ed., Genium Publishing, Schenectady, N.Y., p. 297.

28. d. µg/L

Refer to Smith, R.-K. (1995) *Water and Wastewater Laboratory Techniques.* Water Environ. Fed., Alexandria, Va., p. 166.

29. b. milligrams per liter as $CaCO_3$

Refer to U.S. Environmental Protection Agency (1983) *Methods for Chemical Analysis of Water and Wastewater.* EPA-600/4-79-020 (revised March 1983), Environ. Monit. Support Lab., Cincinnati, Ohio, p. 130.1-1.

30. b. 5 minutes

Refer to U.S. Environmental Protection Agency (1983) *Methods for Chemical Analysis of Water and Wastewater.* EPA-600/4-79-020 (revised March 1983), Environ. Monit. Support Lab., Cincinnati, Ohio, p. 130.2-3.

31. b. 15 mL

Refer to U.S. Environmental Protection Agency (1983) *Methods for Chemical Analysis of Water and Wastewater.* EPA-600/4-79-020 (revised March 1983), Environ. Monit. Support Lab., Cincinnati, Ohio, p. 130.2-3.

32. b. Collect in glass or plastic and cool to 4 °C with a maximum hold time of 48 hours

Refer to U.S. Code of Federal Regulations (1999) *Guidelines Establishing Test Procedures for Analysis of Pollutants Under Clean Water Act.* 40 CFR, Part 136, Washington, D.C., Table II.

33. d. KH_2PO_4/Na_2HPO_4

Refer to American Public Health Association; American Water Works Association; and Water Environment Federation (1992) *Standard Methods for the Examination of Water and Wastewater.* 18th Ed., Washington, D.C., p. 4-67.

34. c. That tends to inhibit changes in pH when an acid or base is added

Refer to California State University (1992) *Operation of Wastewater Treatment Plants.* 4th Ed., Sacramento, Calif., p. 699.

35. d. Molarity

Refer to American Public Health Association; American Water Works Association; and Water Environment Federation (1992) *Standard Methods for the Examination of Water and Wastewater.* 18th Ed., Washington, D.C., p. 1-25.

36. c. Hydrogen

Refer to American Public Health Association; American Water Works Association; and Water Environment Federation (1992) *Standard Methods for the Examination of Water and Wastewater.* 18th Ed., Washington, D.C., inside front cover.

37. d. Ammonia

Refer to Smith, R.-K. (1999) *Handbook of Environmental Analysis.* 4th Ed., Genium Publishing, Schenectady, N.Y., p. 16.

SAFETY

Questions

1. When diluting acid with water

    a. Add water to acid
    b. Add acid to water
    c. Do not mix acids and water
    d. Stand a considerable distance away from the bench in case of explosion

2. The correct procedure for removing contaminated gloves is to

    a. Wash the gloves before removal and then wash your hands after removing the gloves
    b. Wash your hands after removing the gloves
    c. Wash the gloves before removing them
    d. Discard gloves and wash your hands if you touched the outside of the gloves

3. Which of the following is NOT part of the correct procedure for inserting glass tubing to a rubber stopper?

    a. Wear gloves to protect your hands in the event of broken tubing
    b. Hold the glass tubing as close to the rubber stopper as possible
    c. Lubricate the glass with water
    d. Lubricate the glass with silicone grease

4. Dispose of cracked or broken glassware

    a. In the general laboratory waste bins
    b. After washing to remove gross contaminants

        c.   In waste bins marked specifically for glass disposal
        d.   In a covered container

5.   Laboratory fume hoods are used primarily for

        a.   Hazardous chemical storage
        b.   Containment and venting during use of hazardous chemicals
        c.   Odor removal
        d.   General chemical storage

6.   Flammable solvents should be stored

        a.   In properly vented cabinets approved by NFPA
        b.   In the fume hood
        c.   In a refrigerator approved by the National Refrigeration Association
        d.   In the laboratory storage room with other chemicals

7.   Routine laboratory personal protective equipment includes

        a.   Safety glasses, gloves, and aprons
        b.   Steel-toed boots and hard hats
        c.   Self-contained breathing apparatus
        d.   Disposable fire-retardant clothing

8.   MSDS is the recognized acronym for

        a.   Material storage data system
        b.   Management safety data system
        c.   Maintenance and security data system
        d.   Material safety data sheets

9. Chipped glassware

    a. Should never be used
    b. Should be disposed of immediately
    c. May be used if you wear gloves
    d. May be used if the chipped area is fire polished to remove sharp edges

10. Which of the following is required by law in the United States?

    a. An MSDS must be provided with all chemical purchases
    b. Employees must have access to MSDSs for all chemicals with which they may potentially have contact
    c. Treatment facilities must conduct neighborhood training sessions about how to read MSDSs
    d. Chemical supply houses must implement MSDS training

11. Compressed gas cylinders should be stored

    a. Outside the laboratory building
    b. Capped and secured to prevent tipping or rolling
    c. Horizontally to prevent tipping
    d. In a chemical storage area

12. If you have a hexane fire, which type of extinguisher should you use?

    a. Class A
    b. Class B
    c. Class C
    d. Class D

13. In the event of a chemical splash in the eyes, gently flush the eyes for a minimum of

   a. 5 minutes
   b. 10 minutes
   c. 15 minutes
   d. 30 minutes

14. Which of the following most closely summarizes storage recommendations for acids and bases?

   a. Store in well-ventilated cabinets
   b. Separate acids from bases and store in well-ventilated areas away from volatile organic and oxidizable materials
   c. Store with solvents whenever possible
   d. Store only in specially designed cabinets designated for hazardous materials

15. One waterborne disease believed to be caused by a virus is

   a. Cholera
   b. Dysentery
   c. Hepatitis
   d. Tuberculosis

16. Which of the following should NOT be used for pipeting samples in an environmental laboratory?

   a. Three-way bulb
   b. Standard bulb

c. Mouth
d. Digital pipet

17. To provide immediate response to a chemical burn, what should be used?

    a. NaHCO₃
    b. Salve or antiseptic ointment
    c. Large quantities of water
    d. Saline solution

18. Which of the following are corrosive chemicals?

    a. H_2SO_4 and NaOH
    b. KOH and petroleum ether
    c. CS_2 and CCl_4
    d. H_2O_2 and EDTA

19. If you have an electrical fire what type of extinguisher would you use?

    a. Class A
    b. Class B
    c. Class C
    d. Class D

20. Which poisonous gas is commonly found in raw wastewater and has a rotten egg odor?

    a. CO_2
    b. CH_4
    c. H_2S
    d. Chloride

21. Flammable solvents are liquids with

    a. A flash point higher than 60 °C
    b. A strong odor of iodine
    c. A flash point lower than 60 °C
    d. A strong odor of sulfide

22. Properly functioning laboratory fume hoods should have what air flow when the sash is partially closed?

    a. 10 m/min
    b. 30 m/min
    c. 60 m/min
    d. 90 m/min

23. What type of glove is suitable to wear when performing solvent extractions?

    a. Cotton
    b. Nylon
    c. Nitrile
    d. Latex

Answers

1. b. Add acid to water

    Refer to American Public Health Association; American Water Works Association; and Water Environment Federation (1992) *Standard Methods for the Examination of Water and Wastewater*. 18th Ed., Washington, D.C., p. 1-26.

2. a. Wash the gloves before removal and then wash your hands after removing the gloves

    Refer to California State University (1992) *Operation of Wastewater Treatment Plants*. 4th Ed., Sacramento, Calif., p. 474.

3. d. Lubricate the glass with silicone grease

    Refer to California State University (1992) *Operation of Wastewater Treatment Plants*. 4th Ed., Sacramento, Calif., p. 475.

4. c. In waste bins marked specifically for glass disposal

    Refer to California State (1992) *Operation of Wastewater Treatment Plants*. 4th Ed., Sacramento, Calif., p. 476.

5. b. Containment and venting during use of hazardous chemicals

    Refer to American Public Health Association; American Water Works Association; and Water Environment Federation (1992) Standard Methods for the Examination of Water and Wastewater. 18th Ed., Washington, D.C., p. 1-36.

6. a. In properly vented cabinets approved by NFPA

    Refer to American Public Health Association; American Water Works Association; and Water Environment Federation (1992) *Standard Methods for the Examination of Water and Wastewater.* 18th Ed., Washington, D.C., p. 1-36.

7. a. Safety glasses, gloves, and aprons

    Refer to California State University (1992) *Operation of Wastewater Treatment Plants.* 4th Ed., Sacramento, Calif., p. 473.

8. d. Material safety data sheets

    Refer to California State University (1992) *Operation of Wastewater Treatment Plants.* 4th Ed., Sacramento, Calif., p. 472.

9. d. May be used if the chipped area is fire polished to remove sharp edges

    Refer to California State University (1992) *Operation of Wastewater Treatment Plants.* 4th Ed., Sacramento, Calif., p. 475.

10. b. Employees must have access to MSDSs for all chemicals with which they may potentially have contact

    Refer to Smith, R.-K. (1999) *Handbook of Environmental Analysis.* 4th Ed., Genium Publishing, Schenectady, N.Y., p. 1-35.

11. b. Capped and secured to prevent tipping or rolling

    Refer to California State University (1992) *Operation of Wastewater Treatment Plants.* 4th Ed., Sacramento, Calif., p. 474.

12. b. Class B

    Refer to California State University (1992) *Operation of Wastewater Treatment Plants.* 4th Ed., Sacramento, Calif., p. 476.

13. c. 15 minutes

    Refer to American Public Health Association; American Water Works Association; and Water Environment Federation (1992) *Standard Methods for the Examination of Water and Wastewater.* 18th Ed., Washington, D.C. p. 1-35.

14. b. Separate acids from bases and store in well-ventilated areas away from volatile organic and oxidizable materials

    Refer to American Public Health Association; American Water Works Association; and Water Environment Federation (1992) *Standard Methods for the Examination of Water and Wastewater.* 18th Ed., Washington, D.C., p. 1-37.

15. c. Hepatitis

    Refer to California State University (1992) *Operation of Wastewater Treatment Plants.* 4th Ed., Sacramento, Calif., p. 474.

16. c. Mouth

    Refer to American Public Health Association; American Water Works Association; and Water Environment Federation (1992) *Standard Methods for the Examination of Water and Wastewater.* 18th Ed., Washington, D.C., p. 1-38.

17. c. Large quantities of water

Refer to California State University (1992) *Operation of Wastewater Treatment Plants*. 4th Ed., Sacramento, Calif., p. 475.

18. a. H_2SO_4 and NaOH

Refer to California State University (1992) *Operation of Wastewater Treatment Plants*. 4th Ed., Sacramento, Calif., p. 473.

19. c. Class C

This class is used for all electrical fires and in areas where live electricity is present.

Refer to California State University (1992) *Operation of Wastewater Treatment Plants*. 4th Ed., Sacramento, Calif., p. 476.

20. c. H_2S

Refer to California State University (1992) *Operation of Wastewater Treatment Plants*. 4th Ed., Sacramento, Calif., p. 473.

21. c. A flash point lower than 60 °C

Refer to Smith, R.-K. (1999) *Handbook of Environmental Analysis*. 4th Ed., Genium Publishing, Schenectady, N.Y., p. 436.

22. b. 30 m/min

Refer to American Public Health Association; American Water Works Association; and Water Environment Federation (1992) *Standard Methods for the Examination of Water and Wastewater.* 18th Ed., Washington, D.C., p. 1-36.

23. c. Nitrile

Refer to American Public Health Association; American Water Works Association; and Water Environment Federation (1992) *Standard Methods for the Examination of Water and Wastewater.* 18th Ed., Washington, D.C., p. 1-36.

LABORATORY APPARATUS/REAGENTS/TECHNIQUES

Questions

1. NaOH should be stored in a

    a. Polyethylene bottle
    b. Graduated cylinder with a stopper
    c. Borosilicate bottle
    d. Beaker covered with parafilm

2. Which of the following would not be considered volumetric glassware?

    a. Graduated cylinders
    b. Bottles
    c. Pipets
    d. Burets

3. TC means

    a. To concentrate
    b. To collect
    c. To contain
    d. To correct

4. Fritted glassware is

    a. Porous
    b. Nonporous
    c. Plastic
    d. Nonplastic

5. What are used to deliver accurate volumes and range in size from 0.1 to 100 mL?

   a. Beakers
   b. Graduated cylinders
   c. Pipets
   d. Measuring cylinders

6. Beakers are used for

   a. Measuring volumes
   b. Mixing chemicals
   c. Delivering accurate volumes
   d. Delivering and measuring accurate volumes used in titrations

7. Pipets used to deliver one specific volume are

   a. Graduated pipets
   b. Volumetric pipets
   c. Serological pipets
   d. Bacteriological pipets

8. Which pipet should never be "blown out"?

   a. Graduated pipet
   b. Volumetric pipet
   c. Serological pipet
   d. Milk pipet

9. Pipets that require the small amount of liquid remaining in the tip to be "blown out" are identified by

    a. The letters TC near the top of the pipet
    b. The letters TD near the top of the pipet
    c. A frosted band near the top of the pipet
    d. The letters TB near the top of the pipet

10. Accurately calibrated glassware for precise measurements of volume is known as

    a. Graduated glassware
    b. Volumetric glassware
    c. Borosilicate glassware
    d. Corning glassware

11. To correctly read volumetric glassware, which part of the meniscus should be tangent to the calibration mark when viewed at eye level?

    a. Bottom
    b. Middle
    c. Top
    d. Any of the above as long as you are consistent

12. The correct procedure for delivering a solution from a volumetric pipet is to hold the pipet vertically and

    a, Keep the tip of the pipet in contact with the receiving vessel for several seconds after the free flow of liquid has stopped
    b. "Blow out" the pipet after the free flow of liquid has stopped
    c. Shake any droplets from the tip of the pipet into the receiving

vessel after the free flow of liquid has stopped

d. "Blow out" the pipet during the delivery of the liquid

13. All glassware used to deliver liquids must be absolutely clean so that, as the vessel is emptied, the film of liquid coating the inside of the delivery vessel never breaks at any point. This is known as

   a. Draining action
   b. Film action
   c. Sheeting action
   d. Delivery action

14. The correct procedure to prepare a buret for use just after cleaning with soap and water is to

   a. Dry the buret with compressed air
   b. Dry the buret with any inert gas
   c. Rinse the buret with acetone
   d. Rinse the buret two to three times with a small volume of the solution with which it is to be filled

15. Laboratory glassware that meets U.S. specifications for certified glassware is designated as

   a. Class A
   b. Class 1
   c. Analytical grade
   d. TC or TD

16. Laboratory chemicals and reagents are available in a variety of grades of purity. Which grade of chemical is recommended for general laboratory use?

    a. Technical grade
    b. Analytical reagent grade
    c. Pure grade
    d. Ultrapure grade

17. All anhydrous reagent chemicals used for making standard calibration solutions and titrants must be prepared before use by

    a. Drying overnight in a desiccator
    b. Drying in an oven at 305 to 310 °C for at least 1 to 2 hours and desiccating overnight
    c. Drying in an oven at 305 to 310 °C for at least 1 to 2 hours or overnight and cooling to room temperature in a desiccator
    d. Drying in an oven at 105 to 110 °C for at least 1 to 2 hours or overnight and cooling to room temperature in a desiccator

18. In additive solutions (a + b), the first number *a* refers to the volume of concentrated reagent. The second number *b* refers to

    a. The volume of 95% ethyl alcohol required for dilution
    b. The volume of distilled water required for dilution
    c. The volume of 50/50 HCl required for dilution
    d. The volume of concentrated H_2SO_4 required for dilution

19. The proper filter used in the TSS analysis is

    a. A 47-mm filter disk
    b. A glass fiber filter with organic binder

c. A glass fiber filter without organic binder
d. A < 0.2-μm filter disk

20. If you remove too much reagent from the bottle when making a solution, you should

    a. Return excess reagent to the bottle
    b. Pour excess reagent down the drain and flush well
    c. Save excess reagent in a clean container
    d. Dispose of excess reagent according to the label and MSDS directions

21. Laboratory-grade distilled water is prepared by

    a. Distilling pretreated water in a borosilicate glass, fused quartz, tin, or titanium still
    b. Filtering water through several beds of resin cartridges
    c. Reverse osmosis
    d. Ion exchange

22. When reading a buret, the eye must be level to the meniscus of the liquid to eliminate which of the following?

    a. Parallax errors
    b. Refractive index changes
    c. Calibration errors
    d. Reflection effects

23. Which flask gives you the most accurate measurement?

    a. Volumetric flask
    b. Erlenmeyer flask

c. Kjeldahl flask

d. Filtering flask

24. A desiccator is used to

   a. Evaporate moisture from samples
   b. Filter SS
   c. Prevent moisture from entering the sample
   d. Weigh solids to the nearest milligram

25. Which type of tubing is most chemically inert?

   a. Teflon
   b. Polypropylene
   c. Tygon
   d. Copper

26. On which principle is spectrophotometry based?

   a. Boyles' law
   b. Dumas' law
   c. Beer's law
   d. Houle's law

27. An Imhoff cone is used in the measurement of

   a. SS in milligrams per liter
   b. Settleable solids in milligrams per liter
   c. Settleable solids in milliliters per liter
   d. Dissolved solids in milligrams per liter

28. When establishing a standard curve for spectrophotometric analysis, how should standards and blanks be treated?

   a. As samples, using complete method of analysis
   b. Method of analysis is irrelevant
   c. With special care; no need for preliminary treatment
   d. As special samples apart from regular analysis

29. A microburet is a buret that is graduated in

   a. Thousandths of a milliliter
   b. Hundredths of a milliliter
   c. Tenths of a milliliter
   d. Milliliters

30. Changes in ambient temperature result in changes in the capacity of volumetric glassware. Most volumetric glassware is calibrated to a specific volume at a specific temperature, and solutions should be measured at this temperature. This temperature is

   a. 20 °C
   b. 22 °C
   c. 25 °C
   d. 28 °C

31. What do the reagent-grade water specifications type I, type II, and type III indicate?

   a. Type I is the lowest reagent grade
   b. Type III is the highest reagent grade

c. Type II is bacteria free
d. Type I has no detectable concentration of the compound to be analyzed

32. The indicators of reagent-grade water purity are bacteria, pH, particulate matter, resistivity, conductivity, and

   a. Iron
   b. SiO_2
   c. Turbidity
   d. Alkalinity

33. Which of the following is not part of an atomic absorption spectrophotometer?

   a. Column
   b. Nebulizer
   c. Detector
   d. Readout

34. To clean inorganic compounds from laboratory glassware, it is strongly recommended that which of the following is used?

   a. Soap and water
   b. Just water
   c. Strong base
   d. Strong acid

35. At the endpoint of a titration, adding one drop of sample to the titrated solution will

   a. Reverse the color change
   b. Reverse the color change briefly

c. Enhance the color change
d. Do nothing

36. What is the temperature at which all volumetric glassware is calibrated?

a. 4 °C
b. 20 °C
c. 100 °C
d. Room temperature

37. What is the name of the procedure used to check the strength of a titrant?

a. Titration
b. Standardization
c. Concentration analysis
d. Normality determination

Answers

1. a. Polyethylene bottle

   b. and d. are improper storage containers, and the use of a borosilicate bottle could lead to lid freeze.

   Refer to U.S. Environmental Protection Agency (1979) *Handbook for Analytical Quality Control in Water and Wastewater Laboratories.* EPA-600/4-79-019, Environ. Monit. Support Lab., Cincinnati, Ohio, p. 4-2.

2. b. Bottles

   Refer to U.S. Environmental Protection Agency (1979) *Handbook for Analytical Quality Control in Water and Wastewater Laboratories.* EPA-600/4-79-019, Environ. Monit. Support Lab., Cincinnati, Ohio, p. 4-2.

3. c. To contain

   Refer to U.S. Environmental Protection Agency (1979) *Handbook for Analytical Quality Control in Water and Wastewater Laboratories.* EPA-600/4-79-019, Environ. Monit. Support Lab., Cincinnati, Ohio, p. 4-3.

4. a. Porous

   Refer to U.S. Environmental Protection Agency (1979) *Handbook for Analytical Quality Control in Water and Wastewater Laboratories.* EPA-600/4-79-019, Environ. Monit. Support Lab., Cincinnati, Ohio, p. 4-8.

5. c. Pipets

Refer to U.S. Environmental Protection Agency (1979) *Handbook for Analytical Quality Control in Water and Wastewater Laboratories*. EPA-600/4-79-019, Environ. Monit. Support Lab., Cincinnati, Ohio, p. 4-3.

6. b. Mixing chemicals
Beakers are not calibrated accurately enough to deliver accurate volumes.

Refer to U.S. Environmental Protection Agency (1979) *Handbook for Analytical Quality Control in Water and Wastewater Laboratories*. EPA-600/4-79-019, Environ. Monit. Support Lab., Cincinnati, Ohio, p. 4-2.

7. b. Volumetric pipets
Volumetric pipets are calibrated to deliver a fixed volume.

Refer to U.S. Environmental Protection Agency (1979) *Handbook for Analytical Quality Control in Water and Wastewater Laboratories*. EPA-600/4-79-019, Environ. Monit. Support Lab., Cincinnati, Ohio, p. 4-3.

8. b. Volumetric pipet

Refer to U.S. Environmental Protection Agency (1979) *Handbook for Analytical Quality Control in Water and Wastewater Laboratories*. EPA-600/4-79-019, Environ. Monit. Support Lab., Cincinnati, Ohio, p. 4-3.

9. c. A frosted band near the top of the pipet
TC and TD indicate to contain and to deliver types of volumetric glassware.

Refer to U.S. Environmental Protection Agency (1979) *Handbook for Analytical Quality Control in Water and Wastewater Laboratories.* EPA-600/4-79-019, Environ. Monit. Support Lab., Cincinnati, Ohio, p. 4-3.

10. b. Volumetric glassware

Refer to U.S. Environmental Protection Agency (1979) *Handbook for Analytical Quality Control in Water and Wastewater Laboratories.* EPA-600/4-79-019, Environ. Monit. Support Lab., Cincinnati, Ohio, p. 4-2.

11. a. Bottom

Volumetric glassware is only accurate if filled to the point where the bottom of the meniscus just touches the top of the calibration mark. Under- or overfilling causes errors in the measurement.

Refer to U.S. Environmental Protection Agency (1979) *Handbook for Analytical Quality Control in Water and Wastewater Laboratories.* EPA-600/4-79-019, Environ. Monit. Support Lab., Cincinnati, Ohio, p. 4-2.

12. a. Keep the tip of the pipet in contact with the receiving vessel for several seconds after the free flow of liquid has stopped

Refer to U.S. Environmental Protection Agency (1979) *Handbook for Analytical Quality Control in Water and Wastewater Laboratories.* EPA-600/4-79-019, Environ. Monit. Support Lab., Cincinnati, Ohio, p. 4-2.

13. c. Sheeting action

Refer to U.S. Environmental Protection Agency (1979) *Handbook for Analytical Quality Control in Water and Wastewater Laboratories.* EPA-600/4-79-019, Environ. Monit. Support Lab., Cincinnati, Ohio, p. 4-3.

14. d. Rinse the buret two to three times with a small volume of the solution with which it is to be filled

Drying the buret with a gas or other solution/chemical other than the titrant may cause contamination.

Refer to U.S. Environmental Protection Agency (1979) *Handbook for Analytical Quality Control in Water and Wastewater Laboratories.* EPA-600/4-79-019, Environ. Monit. Support Lab., Cincinnati, Ohio, p. 4-3.

15. a. Class A

Refer to U.S. Environmental Protection Agency (1979) *Handbook for Analytical Quality Control in Water and Wastewater Laboratories.* EPA-600/4-79-019, Environ. Monit. Support Lab., Cincinnati, Ohio, p. 4-4.

16. b. Analytical reagent grade

Refer to U.S. Environmental Protection Agency (1979) *Handbook for Analytical Quality Control in Water and Wastewater Laboratories.* EPA-600/4-79-019, Environ. Monit. Support Lab., Cincinnati, Ohio, p. 5-1.

17. d. Drying in an oven at 105 to 110 °C for at least 1 to 2 hours or overnight, and cooling to room temperature in a desiccator

Refer to American Public Health Association; American Water Works Association; and Water Environment Federation (1992) *Standard Methods for the Examination of Water and Wastewater.* 18th Ed., Washington, D.C., p. 1-25.

18. b. The volume of distilled water required for dilution

Refer to Smith, R.-K. (1995) *Water and Wastewater Laboratory Techniques.* Water Environ. Fed., Alexandria, Va., p. 106.

19. c. A glass fiber filter without organic binder

Refer to American Public Health Association; American Water Works Association; and Water Environment Federation (1992) *Standard Methods for the Examination of Water and Wastewater.* 18th Ed., Washington, D.C., p. 2-55.

20. d. Dispose of excess reagent according to the label and MSDS directions

Refer to American Public Health Association; American Water Works Association; and Water Environment Federation (1992) *Standard Methods for the Examination of Water and Wastewater.* 18th Ed., Washington, D.C., p. 1-41.

21. a. Distilling pretreated water in a borosilicate glass, fused quartz, tin, or titanium still

Refer to American Public Health Association; American Water Works Association; and Water Environment Federation (1992) *Standard Methods for the Examination of Water and Wastewater.* 18th Ed., Washington, D.C., p. 1-33.

22. a. Parallax errors

Refer to Smith, R.-K. (1995) *Water and Wastewater Laboratory Techniques.* Water Environ. Fed., Alexandria, Va., p. 86.

23. a. Volumetric flask

Refer to Smith, R.-K. (1995) *Water and Wastewater Laboratory Techniques.* Water Environ. Fed., Alexandria, Va., p. 80.

24. c. Prevent moisture from entering the sample

Refer to Smith, R.-K. (1995) *Water and Wastewater Laboratory Techniques.* Water Environ. Fed., Alexandria, Va., p. 152.

25. a. Teflon

Teflon resists acids, alkalis, organic solvents, and synthetic lubricants.

Refer to Smith, R.-K. (1995) *Water and Wastewater Laboratory Techniques.* Water Environ. Fed., Alexandria, Va., p. 55.

26. c. Beer's law

The concentration of a light-absorbing colored solution is directly proportional to the absorbance over a given range of concentrations.

Refer to American Public Health Association; American Water Works Association; and Water Environment Federation (1992) *Standard Methods for the Examination of Water and Wastewater.* 18th Ed., Washington, D.C., p. 1-28.

27. c. Settleable solids in milliliters per liter

Refer to American Public Health Association; American Water Works Association; and Water Environment Federation (1992) *Standard Methods for the Examination of Water and Wastewater.* 18th Ed., Washington, D.C., p. 2-57.

28. a. As samples, using complete method of analysis

Refer to American Public Health Association; American Water Works Association; and Water Environment Federation (1992) *Standard Methods for the Examination of Water and Wastewater.* 18th Ed., Washington, D.C., p. 1-5.

29. b. Hundredths of a milliliter

Refer to U.S. Environmental Protection Agency (1979) *Handbook for Analytical Quality Control in Water and Wastewater Laboratories.* EPA-600/4-79-019, Environ. Monit. Support Lab., Cincinnati, Ohio, p. 4-3.

30. a. 20 °C

1000 mL of water increases in volume approximately 0.20 mL per 1 °C rise in ambient temperature.

Refer to U.S. Environmental Protection Agency (1979) *Handbook for Analytical Quality Control in Water and Wastewater Laboratories.* EPA-600/4-79-019, Environ. Monit. Support Lab., Cincinnati, Ohio, p. 4-3.

31. d. Type I has no detectable concentration of the compound to be analyzed

Refer to American Public Health Association; American Water Works Association; and Water Environment Federation (1992) *Standard Methods for the Examination of Water and Wastewater.* 18th Ed., Washington, D.C., p. 1-32.

32. b. SiO_2

Refer to American Public Health Association; American Water Works Association; and Water Environment Federation (1992) *Standard Methods for the Examination of Water and Wastewater.* 18th Ed., Washington, D.C., p. 1-32.

33. a. Column

Refer to American Public Health Association; American Water Works Association; and Water Environment Federation (1992) *Standard Methods for the Examination of Water and Wastewater.* 18th Ed., Washington, D.C., p. 3-11.

34. d. Strong acid

Refer to American Public Health Association; American Water Works Association; and Water Environment Federation (1992) *Standard Methods for the Examination of Water and Wastewater.* 18th Ed., Washington, D.C., p. 1-24.

35. a. Reverse the color change

Refer to Smith, R.-K. (1995) *Water and Wastewater Laboratory Techniques.* Water Environ. Fed., Alexandria, Va., p. 163.

36. b. 20 °C

Refer to Smith, R.-K. (1995) *Water and Wastewater Laboratory Techniques.* Water Environ. Fed., Alexandria, Va., p. 84.

37. b. Standardization

Refer to Smith, R.-K. (1995) *Water and Wastewater Laboratory Techniques.* Water Environ. Fed., Alexandria, Va., p. 109.

MATHEMATICS

Questions

1. One liter of 1 M NaCl solution contains how many grams of NaCl?

    a. 22 g
    b. 29 g
    c. 35 g
    d. 58 g

2. A laboratory technician wishes to heat a sample at 103 °C, but the oven is calibrated in degrees Fahrenheit. The oven should be set to read

    a. 128 °F
    b. 153 °F
    c. 158 °F
    d. 217 °F

3. The temperature reading was 15 °C. What is the temperature in degrees Fahrenheit?

    a. 26 °F
    b. 40 °F
    c. 59 °F
    d. 85 °F

4. What degrees Celsius is equivalent to 61 °F?

    a. 16 °C
    b. 38 °C

c. 61 °C
d. 76 °C

5. What temperature in degrees Celsius is equivalent to 95 °F?

    a. 35 °C
    b. 37 °C
    c. 63 °C
    d. 127 °C

6. The temperature of a waste stream is 85 °F, what is the temperature in degrees Celsius?

    a. 29 °C
    b. 53 °C
    c. 85 °C
    d. 95 °C

7. What temperature in degrees Fahrenheit is equivalent to 31 °C?

    a. 85 °F
    b. 88 °F
    c. 89 °F
    d. 90 °F

8. The conversion factor between milligrams per liter and percent is

    a. 1 000 mg/L
    b. 10 000 mg/L

c. 100 000 mg/L
d. 1 000 000 mg/L

9. Find the standard deviation of the variance $v = 139.82$.

   a. 11.8
   b. 16.8
   c. 69.5
   d. 277.8

10. Given 100 ppm, 250 ppm, 150 ppm, 105 ppm, 208 ppm, and 188 ppm as data, what is the range?

    a. 100
    b. 150
    c. 153.5
    d. 154

11. Given 160, 155, 160, 160, 180, 165, 155, 170, 160, 165, 155, 150, 145, and 160 as results, what is the arithmetic mean?

    a. 35
    b. 160
    c. 165
    d. 180

12. Given 160 mg/L, 155 mg/L, 160 mg/L, 160 mg/L, 180 mg/L, 165 mg/L, 155 mg/L, 170 mg/L, 160 mg/L, 165 mg/L, 155 mg/L, 150 mg/L, 145 mg/L, and 160 mg/L as values, what is the range?

    a. 35

b. 160
c. 165
d. 180

13. Given 220, 7200, 230, 300, 240, 270, 240, 250, 240, and 260 as data, determine the median value.

    a. 240
    b. 245
    c. 300
    d. 881.8

14. 5.31 mL of 0.12 M $K_2Cr_2O_7$ solution contains how many milligrams of $K_2Cr_2O_7$?

    a. 44 mg
    b. 148 mg
    c. 187 mg
    d. 296 mg

15. 4.10 g of $K_2Cr_2O_7$ is added to 750 mL of 0.141 M $K_2Cr_2O_7$. What is the final concentration?

    a. 0.014 M
    b. 0.019 M
    c. 0.141 M
    d. 0.160 M

16. 10.71 mL of 0.020 N H_2SO_4 contains how many milligrams of H_2SO_4?

    a. 5.25 mg
    (b.) 10.5 mg

c. 21.0 mg
d. 42.0 mg

17. Concentrated HCl is 37% by weight in water. How many grams of concentrated acid must be diluted to 1000 mL to give a 1.00 M solution of HCl?

   a. 24.3 g
   b. 48.6 g
   ● c. 97.3 g
   d. 195 g

18. How many milliliters of a 0.75 M $K_2Cr_2O_7$ solution should be diluted to give 100 mL of 0.025 M concentration $K_2Cr_2O_7$ solution?

   a. 0.33 mL
   ● b. 3.33 mL
   c. 33.3 mL
   d. 333 mL

19. Concentrated HNO_3 is 70% by weight. How many grams of concentrated HNO_3 must be diluted to 1000 mL to give a 0.33 M solution of HNO_3?

   a. 21.1 g
   ● b. 29.7 g
   c. 47.1 g
   d. 63.3 g

20. Water density at 20 °C is 0.998 203 g/mL. A drum is filled with a water solution at 20 °C that has a density of 1.120 1 g/mL. How much solid is dissolved in the water?

    a. 121 mg/L
    b. 1218 mg/L
    c. 12 189 mg/L
    d. 121 897 mg/L

21. 356 mg additional potassium hydrogen phthalate ($KHC_8H_4O_4$) is added to 1000 mL of 0.050 M potassium hydrogen phthalate (MW = 204). What is the final concentration?

    a. 0.048 3 M
    b. 0.050 0 M
    c. 0.051 7 M
    d. 0.100 0 M

22. To what is 35 °C equivalent in degrees Fahrenheit?

    a. 57 °F
    b. 63 °F
    c. 85 °F
    d. 95 °F

23. To convert Fahrenheit to Celsius

    a. Add 32 and multiply by 1.8
    b. Subtract 32 and multiply by 0.555 6
    c. Subtract 32 and multiply by 1.8
    d. None of the above

24. Wastewater entering the plant has a BOD of 300 mg/L and the plant effluent has a BOD of 30 mg/L. What is the percent removal of BOD through the plant?

    a. 10%
    b. 60%
    c. 80%
    d. 90%

25. Raw influent has a BOD of 180 mg/L. Final effluent has a BOD of 12 mg/L. What is the percent removal?

    a. 78%
    b. 84%
    c. 93%
    d. 97%

26. Calculate the percent removal of BOD if the raw BOD is 250 mg/L and the final effluent BOD is 20 mg/L.

    a. 23%
    b. 70%
    c. 82%
    d. 92%

27. From the following data, what is the BOD?

    Initial DO = 8.2 mg/L,
    Final DO = 4.2 mg/L,
    Sample Volume = 6 mL, and
    Bottle Volume = 300 mL.

a. 200 mg/L
b. 210 mg/L
c. 400 mg/L
d. 420 mg/L

28. A sample of sludge is placed in an evaporating dish, weighed, dried, cooled, weighed, ignited in a muffle furnace, cooled, and weighed. What is the percent total solids of the sludge?

Data:
Weight of Empty Dish = 37.25 g,
Weight of Dish After Filling = 97.25 g,
Weight of Dish After Drying = 40.50 g, and
Weight of Dish After Ignition = 39.50 g.

a. 0.3%
b. 1.4%
c. 2.7%
d. 5.4%

29. Compute the SS concentration based on the following data:

Initial Weight of Filter Disk = 0.369 g,
Volume of Filtered Sample = 50 mL, and
Weight of Filter Disk and Dried Residue = 0.519 g.

a. 30 mg/L
b. 150 mg/L
c. 1500 mg/L
d. 3000 mg/L

30. From the following data, calculate the coliform density.

Sample used, mL	Colonies counted
5	33
10	75
15	92

    a. 330 CFU/100 mL
    b. 627 CFU/100 mL
    c. 660 CFU/100 mL
    d. 920 CFU/100 mL

31. Given the following data, calculate the BOD for a sample.

    Bottle volume = 300 mL

Sample, mL	Initial DO, mg/L	Final DO, mg/L
Blank	8.5	8.5
4	8.5	7.0
6	8.5	5.8
10	8.5	0.9

    a. 120 mg/L
    b. 124 mg/L
    c. 130 mg/L
    d. 135 mg/L

32. Given the following data, calculate the percent total solids.

    Weight of Empty Dish = 20.55 g,
    Weight of Dish and Wet Sample = 52.61 g,

Weight of Dish and Dry Sample = 20.89 g, and

Weight of Dish and Fixed Sample = 20.65 g.

    a.   1.05%

    b.   1.06%

    c.   2.12%

    d.   6.72%

33. Given the following data, calculate the percent volatile solids.

Weight of Empty Dish = 20.55 g,

Weight of Dish and Wet Sample = 52.61 g,

Weight of Dish and Dry Sample = 20.89 g, and

Weight of Dish and Fixed Sample = 20.65 g.

    a.   35.3%

    b.   70.6%

    c.   77.1%

    d.   97.1%

34. An analyst performs a TSS test on 100 mL of raw domestic wastewater. Given the following data, calculate the TSS level for the sample.

Weight of Filter = 0.415 8 g and

Weight of Filter and Residue = 0.428 5 g

    a.   0.250 mg/L

    b.   127 mg/L

    c.   130 mg/L

    d.   205 mg/L

35. What is the geometric mean of the following numbers: 75, 80, 152, 73?

    a. 76
    b. 95
    c. 90
    d. 140

36. If you placed 1 mL of sample into a 99-mL dilution bottle, 1 mL of that into another 99-mL dilution bottle, and then ran 10 mL of that across a membrane filter, what would be the dilution factor?

    a. 1:100
    b. 1:1000
    c. 1:10 000
    d. 1:100 000

    $100 \times 100 \times 10$

37. Given the following data, determine the seeded BOD for the sample below.

    Bottle Volume = 300 mL

    Sample
    Initial DO = 8.0 mg/L
    Final DO = 4.0 mg/L
    Sample Volume = 15 mL
    Volume of Seed in Sample = 1.8 mL

    Seed Control
    Initial DO = 8.1 mg/L
    Final DO = 5.3 mg/L
    Volume of Seed in Control = 6 mL

    a. 24 mg/L
    b. 63 mg/L
    c. 80 mg/L
    d. 107 mg/L

    $$\frac{(8.0 - 4.0) - (8.1 - 5.3)\left(\frac{1.8}{6}\right)}{15}\left(\frac{300}{}\right)$$

38. In the problem above, what is the value of the seed control ratio f?

   a. 0.3
   b. 0.4
   c. 3.0
   d. 3.3

39. If a sample had an initial DO of 8.9 mg/L, a final DO of 6.4 mg/L, and the volume of sample was 9 mL in a 300-mL bottle, what is the BOD?

   a. 70 mg/L
   b. 83 mg/L
   c. 96 mg/L
   d. 109 mg/L

40. How many milliliters of 0.30 M H_3PO_4 solution are needed to make 1 L of 0.12 M solution?

   a. 200 mL
   b. 400 mL
   c. 600 mL
   d. 800 mL

41. A 25-mL sample is filtered for TSS, the weight of the filter is 110.8 mg, and the weight of the filter and residue is 114.9 mg. What is the TSS of this sample?

   a. 144 mg/L
   b. 154 mg/L
   c. 164 mg/L
   d. 174 mg/L

42. If a sample is analyzed for ammonia and has a reading of 0.83 mg/L, calculate the accuracy for a 2.00-mg/L matrix spike of this sample when the final reading is 2.66 mg/L.

   a. 71%
   b. 76%
   c. 83%
   d. 92%

43. How many grams of KCl are required to make 1 L of a 2.33 M solution?

   a. 175 g
   b. 200 g
   c. 226 g
   d. 252 g

Answers

1.   d.   58 g

$$\text{MW of NaCl} = 58 \ \frac{g}{mol} \quad \text{and} \quad 1\,M = \frac{1\,mol}{1\,L}$$

$$\text{NaCl} = \frac{1\,\cancel{mol}}{1\,\cancel{L}}(1\,\cancel{L})(58\,\frac{g}{\cancel{mol}})$$
$$= 58\,g$$

Refer to Smith, R.-K. (1995) *Water and Wastewater Laboratory Techniques.* Water Environ. Fed., Alexandria, Va., p. 191.

2.   d.   217 °F

°F = (°C × 1.8) + 32
°F = (103 °C × 1.8) + 32
    = 217 °F

Refer to California State University (1992) *Operation of Wastewater Treatment Plants.* 4th Ed., Sacramento, Calif., p. 458.

3.   c.   59 °F

°F = (°C × 1.8) + 32
°F = (15 °C × 1.8) + 32
    = 59 °F

Refer to California State University (1992) *Operation of Wastewater Treatment Plants.* 4th Ed., Sacramento, Calif., p. 458.

4. a. 16 °C

°C = (°F − 32) (0.555 6)
°C = (61 °F − 32) (0.555 6)
 = 16 °C

Refer to California State University (1992) *Operation of Wastewater Treatment Plants.* 4th Ed., Sacramento, Calif., p. 458.

5. a. 35 °C

°C = (°F − 32) (0.555 6)
°C = (95 °F − 32) (0.555 6)
 = 35 °C

Refer to California State University (1992) *Operation of Wastewater Treatment Plants.* 4th Ed., Sacramento, Calif., p. 458.

6. a. 29 °C

°C = (°F − 32) (0.555 6)
°C = (85 °F − 32) (0.555 6)
 = 29 °C

Refer to California State University (1992) *Operation of Wastewater Treatment Plants.* 4th Ed., Sacramento, Calif., p. 458.

7. b. 88 °F

$$°F = (°C \times 1.8) + 32$$
$$°F = (31\ °C \times 1.8) + 32$$
$$= 88\ °F$$

Refer to California State University (1992) *Operation of Wastewater Treatment Plants.* 4th Ed., Sacramento, Calif., p. 458.

8. b. 10 000 mg/L

$$1\ L = 1\ 000\ 000\ mg/L$$
$$100\% = 1\ 000\ 000\ mg/L$$
$$1\% = \frac{1\ 000\ 000\ mg/L}{100\%}$$
$$= 10\ 000\ mg/L$$

Refer to Smith, R.-K. (1999) *Handbook of Environmental Analysis.* 4th Ed., Genium Publishing, Schenectady, N.Y., p. 166.

9. a. 11.8

$$\text{Standard Deviation} = \text{Square Root of the Variance}$$
$$\text{Standard Deviation} = \sqrt{139.82}$$
$$= 11.8$$

Refer to U.S. Environmental Protection Agency (1979) *Handbook for Analytical Quality Control in Water and Wastewater Laboratories.* EPA-600/4-79-019, Environ. Monit. Support Lab., Cincinnati, Ohio, p. 7-4.

10. b. 150

   Range = Largest Value – Smallest Value
   Range = 250 – 100
   　　　 = 150

   Refer to U.S. Environmental Protection Agency (1979) *Handbook for Analytical Quality Control in Water and Wastewater Laboratories*. EPA-600/4-79-019, Environ. Monit. Support Lab., Cincinnati, Ohio, p. 7-4.

11. b. 160

   Mean = Sum of Results/Number of Results
   Sum of Results = 160 + 155 + 160 + 160 + 180 + 165 + 155 + 170 +
   　　　　　　　　　160 + 165 + 155 + 150 + 145 + 160 = 2240
   Number of Results = 14
   Mean = 2240/14 = 160

   Refer to U.S. Environmental Protection Agency (1979) *Handbook for Analytical Quality Control in Water and Wastewater Laboratories*. EPA-600/4-79-019, Environ. Monit. Support Lab., Cincinnati, Ohio, p. 7-3.

12. a. 35

   Range = Largest Value – Smallest Value
   Range = 180 – 145 = 35

   Refer to U.S. Environmental Protection Agency (1979) *Handbook for Analytical Quality Control in Water and Wastewater Laboratories*. EPA-600/4-79-019, Environ. Monit. Support Lab., Cincinnati, Ohio, p. 7-3.

13. b. 245

The median is the middle most value in a set of numbers. If there are two values in the middle of the set, the median is the average of those two values.

(1) Rank the numbers in increasing order: 220, 230, 240, 240, 240, 250, 260, 270, 300, 7200

(2) The fifth and sixth numbers are in the middle: 240 and 250

(3) $\dfrac{240 + 250}{2} = 245$

Refer to U.S. Environmental Protection Agency (1979) *Handbook for Analytical Quality Control in Water and Wastewater Laboratories.* EPA-600/4-79-019, Environ. Monit. Support Lab., Cincinnati, Ohio, p. 7-4.

14. c. 187 mg

$$\text{Molarity} = \frac{(\text{Weight of Compound, g})/(\text{Molecular Weight of Compound, g/mol})}{(\text{Volume of Solution, L})}$$

MW of $K_2Cr_2O_7$ = 294 g/mol = 294 mg/mmol

Molarity is expressed as mol/L or mmol/mL

Therefore,

Weight of Compound, mg
= (Molarity, mmol/mL) (Volume of Sample, mL) (MW of Compound, mg/mmol)

Weight of Compound = $5.31\ \text{mL} \left(\dfrac{0.12\ \text{mmol}}{\text{mL}}\right) \times \left(\dfrac{294\ \text{mg}}{\text{mmol}}\right)$
= 187 mg

Refer to Smith, R.-K. (1995) *Water and Wastewater Laboratory Techniques.* Water Environ. Fed., Alexandria, Va., p. 197.

15. d.   0.160 M

There are two approaches to solving this problem. Knowing that the MW of $K_2Cr_2O_7$ = 294 g/mol, or 294 mg/mmol,

Solution 1: $(0.750 \text{ L})(0.141 \frac{\text{mol}}{\text{L}}) = 0.106 \text{ mol } K_2Cr_2O_7$ in initial solution

$(0.106 \text{ mol})(294 \frac{\text{g}}{\text{mol}}) = 31.2 \text{ g}$ in initial solution

$31.2 \text{ g} + 4.10 \text{ g} = 35.2 \text{ g } K_2Cr_2O_7$ in final solution

$\dfrac{35.2 \text{ g}}{294 \frac{\text{g}}{\text{mol}}} = 0.12 \text{ mol}$ in final solution

$\dfrac{0.12 \text{ mol}}{0.750 \text{ L}} = 0.16 \text{ mol/L} = 0.16 \text{ M}$

Solution 2: $\dfrac{4.10 \text{ g}}{294 \frac{\text{g}}{\text{mol}}} = 0.013 \text{ 9 mol}$

$\dfrac{0.0139 \text{ mol}}{0.750 \text{ L}} = 0.018 \text{ 5 mol/L} = 0.018 \text{ 5 M}$

Total $K_2Cr_2O_7$ Concentration $= 0.018 \text{ 5 M} + 0.141 \text{M} = 0.160 \text{ M}$

Refer to Smith, R.-K. (1995) *Water and Wastewater Laboratory Techniques.* Water Environ. Fed., Alexandria, Va., p. 198.

16. b.   10.5 mg

A normal solution is a solution containing 1 g equivalent weight of solute per liter
MW of H_2SO_4 = 98 g/mol or 98 mg/mmol
Equivalent weight of H_2SO_4 = 98/2 equivalents (H_2) = 49 g/mol or 49 mg/mmol
Using this method you can convert between molar and normal concentrations

0.020 N H_2SO_4/2 equivalents (H_2) = 0.010 M H_2SO_4

$10.71 \text{ mL}(0.010 \frac{\text{mmol}}{\text{mL}}) = 0.107 \text{ 1 mmol}$

$0.107 \text{ 1 mmol}(98 \frac{\text{mg}}{\text{mmol}}) = 10.5 \text{ mg}$

Refer to Smith, R.-K. (1995) *Water and Wastewater Laboratory Techniques.* Water Environ. Fed., Alexandria, Va., p. 198.

17. c. 97.3 g

$$1.00\ M = 1.00\ \frac{mol}{L}$$

MW of HCl = 36 g/mol

Therefore, 36 g of HCl is needed in 1000 mL of solution

$$\text{Mass of Concentrated Acid Needed} = \frac{36\ g}{0.37} = 97.3\ g$$

Refer to Smith, R.-K. (1995) *Water and Wastewater Laboratory Techniques.* Water Environ. Fed., Alexandria, Va., p. 198.

18. b. 3.33 mL

Dilution = (Initial Volume) (Initial Molarity) = (Final Volume) (Final Molarity)

Therefore,

Initial Volume (0.75 M) = 100 mL (0.025 M)

$$\text{Initial Volume} = \frac{(100\ mL)(0.025\ M)}{0.75\ M} = 3.33\ mL$$

Refer to Smith, R.-K. (1995) *Water and Wastewater Laboratory Techniques.* Water Environ. Fed., Alexandria, Va., p. 198.

19. b.   29.7 g

MW of HNO$_3$ = 63 g/mol

1.00 L of 0.33 M solution contains 0.33 mol

HNO$_3$ needed = $0.33 \text{ mol} \left(\dfrac{63 \text{ g}}{\text{mol}} \right) = 20.8$ g

Concentrated Acid Needed = $\dfrac{20.8 \text{ g}}{0.70}$
= 29.7 g

Refer to Smith, R.-K. (1995) *Water and Wastewater Laboratory Techniques*. Water Environ. Fed., Alexandria, Va., p. 197.

20. d.   121 897 mg/L

Weight of Solution = 1.120 1 g/mL = 1120.1 g/L = 1 120 100 mg/L
Weight Contribution of Water = 0.998 203 g/mL = 998.203 g/L = 998 203 mg/L
Difference Due to the Solid = 1 120 100 mg/L − 998 203 mg/L = 121 897 mg/L

Refer to Smith, R.-K. (1995) *Water and Wastewater Laboratory Techniques*. Water Environ. Fed., Alexandria, Va., p. 174.

21. c.   0.051 7 M

MW of potassium hydrogen phthalate (KHC$_8$H$_4$O$_4$) = 204 g/mol = 204 mg/mmol

Concentration of additional potassium hydrogen phthalate = $\dfrac{356 \text{ mg}}{204 \dfrac{\text{mg}}{\text{mmol}}} = 1.74$ mmol

$\dfrac{1.74 \text{ mmol}}{1000 \text{ mL}} = 0.001\ 74$ mmol/L $= 0.001\ 74$ M

Final concentration = 0.001 74 M + 0.050 M
= 0.051 7 M

Refer to Smith, R.-K. (1995) *Water and Wastewater Laboratory Techniques.* Water Environ. Fed., Alexandria, Va., p. 197.

22. d. 95 °F

°F = (°C × 1.8) + 32
°F = (35 °C × 1.8) + 32
   = 95 °F

Refer to Smith, R.-K. (1995) *Water and Wastewater Laboratory Techniques.* Water Environ. Fed., Alexandria, Va., p. 91.

23. b. Subtract 32 and multiply by 0.555 6

Refer to Smith, R.-K. (1995) *Water and Wastewater Laboratory Techniques.* Water Environ. Fed., Alexandria, Va., p. 91.

24. d. 90%

$$\text{Percent Removal} = \frac{\text{Total Concentration In, mg/L} - \text{Total Concentration Out, mg/L}}{\text{Total Concentration In, mg/L}} \times 100\%$$

$$= \frac{300 \text{ mg/L} - 30 \text{ mg/L}}{300 \text{ mg/L}} \times 100\%$$

$$= \frac{270 \text{ mg/L}}{300 \text{ mg/L}} \times 100\%$$

$$= 90\%$$

Refer to California State University (1992) *Operation of Wastewater Treatment Plants.* 4th Ed., Sacramento, Calif., p. 129.

25. c. 93%

$$\text{Percent Removal} = \frac{\text{Total Concentration In, mg/L} - \text{Total Concentration Out, mg/L}}{\text{Total Concentration In, mg/L}} \times 100\%$$

$$= \frac{180 \text{ mg/L} - 12 \text{ mg/L}}{180 \text{ mg/L}} \times 100\%$$

$$= \frac{168 \text{ mg/L}}{180 \text{ mg/L}} \times 100\%$$

$$= 93\%$$

Refer to California State University (1992) *Operation of Wastewater Treatment Plants.* 4th Ed., Sacramento, Calif., p. 129.

26. d. 92%

$$\text{Percent Removal} = \frac{\text{Total Concentration In, mg/L} - \text{Total Concentration Out, mg/L}}{\text{Total Concentration In, mg/L}} \times 100\%$$

$$= \frac{250 \text{ mg/L} - 20 \text{ mg/L}}{250 \text{ mg/L}} \times 100\%$$

$$= \frac{230 \text{ mg/L}}{250 \text{ mg/L}} \times 100\%$$

$$= 92\%$$

Refer to California State University (1992) *Operation of Wastewater Treatment Plants.* 4th Ed., Sacramento, Calif., p. 129.

27. a. 200 mg/L

$$\text{BOD} = \frac{(\text{Initial DO, mg/L} - \text{Final DO, mg/L})}{\left(\dfrac{\text{Sample Volume, mL}}{\text{Bottle Volume, mL}}\right)}$$

$$\text{BOD} = \frac{8.2 \text{ mg/L} - 4.2 \text{ mg/L}}{\left(\dfrac{6 \text{ mL}}{300 \text{ mL}}\right)} = \frac{4.0 \text{ mg/L}}{0.02} = 200 \text{ mg/L}$$

Refer to American Public Health Association; American Water Works Association; and Water Environment Federation (1992) *Standard Methods for the Examination of Water and Wastewater.* 18th Ed., Washington, D.C., p. 5-5.

28. d. 5.4%

$$\text{Percent Total Solids} = \frac{\text{Dry Solids Weight, g}}{\text{Wet Solids Weight, g}} \times 100\%$$

Dry Solids Weight = Weight of Dish and Solids After Drying −
Weight of Empty Dish

Wet Solids Weight = Weight of Dish and Solids Before Drying −
Weight of Empty Dish

$$\text{Percent Total Solids} = \frac{(40.50 \text{ g} - 37.25 \text{ g})}{(97.25 \text{ g} - 37.25 \text{ g})} \times 100\%$$

$$= \frac{3.25 \text{ g}}{60.0 \text{ g}} \times 100\% = 5.4\%$$

Refer to American Public Health Association; American Water Works Association; and Water Environment Federation (1992) *Standard Methods for the Examination of Water and Wastewater.* 18th Ed., Washington, D.C., p. 2-58.

29. d. 3000 mg/L

$$\text{TSS} = \frac{(\text{Weight of Filter and Dry Residue, mg} - \text{Weight of Filter, mg})}{\text{Sample Volume, mL}} \times 1000 \frac{\text{mL}}{\text{L}}$$

$$= \frac{[(0.519\ \text{g} - 0.369\ \text{g})(1000\ \frac{\text{mg}}{\text{g}})]}{50\ \text{mL}} \times 1000\ \frac{\text{mL}}{\text{L}}$$

$$= 3000\ \text{mg/L}$$

Refer to American Public Health Association; American Water Works Association; and Water Environment Federation (1992) *Standard Methods for the Examination of Water and Wastewater.* 18th Ed., Washington, D.C., p. 2-56.

30. c. 660 colonies/100 mL

The 10- and 15-mL samples are out of the range for the membrane filter coliform test (20 to 60 CFU/100 mL).

$$\text{Coliform Density} = \frac{\text{Coliform Colonies} \times 100}{\text{Filtered Sample, mL}}$$

$$= \frac{33 \times 100}{5}$$

$$= 660\ \text{CFU/100 mL}$$

Refer to American Public Health Association; American Water Works Association; and Water Environment Federation (1992) *Standard Methods for the Examination of Water and Wastewater.* 18th Ed., Washington, D.C., p. 9-61.

31. d. 135 mg/L

The 4-mL sample did not deplete 2.0 mg/L; and the 10-mL sample depleted to less than 1.0 mg/L. Therefore, only the 6-mL sample can be calculated.

$$\text{BOD} = \frac{(\text{Initial DO, mg/L} - \text{Final DO, mg/L})}{\dfrac{\text{Sample Volume, mL}}{\text{Bottle Volume, mL}}}$$

$$\text{BOD} = \frac{(8.5 \text{ mg/L} - 5.8 \text{ mg/L})}{\dfrac{6 \text{ mL}}{300 \text{ mL}}}$$

$$= \frac{2.7 \text{ mg/L}}{0.02} = 135 \text{ mg/L}$$

Refer to American Public Health Association; American Water Works Association; and Water Environment Federation (1992) *Standard Methods for the Examination of Water and Wastewater*. 18th Ed., Washington, D.C., p. 5-5.

32. b.   1.06%

$$\text{Percent Total Solids} = \frac{\text{Dry Solids Weight, g}}{\text{Wet Solids Weight, g}} \times 100\%$$

Dry Solids Weight = Weight of Dish and Solids After Drying − Weight of Empty Dish

Wet Solids Weight = Weight of Dish and Solids Before Drying − Weight of Empty Dish

$$\text{Percent Total Solids} = \frac{(20.89 \text{ g} - 20.55 \text{ g})}{(52.61 \text{ g} - 20.55 \text{ g})} \times 100\%$$

$$= \frac{0.34 \text{ g}}{32.06 \text{ g}} \times 100\%$$

$$= 1.06\%$$

Refer to American Public Health Association; American Water Works Association; and Water Environment Federation (1992) *Standard Methods for the Examination of Water and Wastewater*. 18th Ed., Washington, D.C., p. 2-58.

33. b.   70.6%

$$\text{Percent Volatile Solids} = \frac{\text{Dry Solids Weight, g} - \text{Ash Solids Weight, g}}{\text{Dry Solids Weight, g}} \times 100\%$$

Dry Solids Weight = Weight of Dish and Solids After Drying − Weight of Empty Dish

Ash Solids Weight = Weight of Dish and Solids After Ignition − Weight of Empty Dish

$$\text{Percent Volatile Solids} = \frac{[(20.89\text{ g} - 20.55\text{ g}) - (20.65\text{ g} - 20.55\text{ g})]}{(20.89\text{ g} - 20.55\text{ g})} \times 100\%$$

$$= \frac{0.34\text{ g} - 0.10\text{ g}}{0.34\text{ g}} \times 100\%$$

$$= 70.6\%$$

Refer to American Public Health Association; American Water Works Association; and Water Environment Federation (1992) *Standard Methods for the Examination of Water and Wastewater*. 18th Ed., Washington, D.C., p. 2-58.

34. b. 127 mg/L

$$\text{TSS} = \frac{(\text{Weight of Filter and Dry Residue, mg} - \text{Weight of Filter, mg})}{\text{Sample Volume, mL}} \times 1000\,\frac{\text{mL}}{\text{L}}$$

$$= \frac{428.5\text{ mg} - 415.8\text{ mg}}{100\text{ mL}} \times 1000\,\frac{\text{mL}}{\text{L}}$$

$$= 127\text{ mg/L}$$

Refer to American Public Health Association; American Water Works Association; and Water Environment Federation (1992) *Standard Methods for the Examination of Water and Wastewater*. 18th Ed., Washington, D.C., p. 2-56.

35. c. 90

$\log^{10} 75 = 1.875\,1$

$\log^{10} 80 = 1.903\,1$

$\log^{10} 152 = 2.181\,8$

$\log^{10} 73 = 1.863\,3$

Sum of log^{10} results = 7.823 3

Average = $\dfrac{7.823\ 3}{4} = 1.955\ 8$

Inverse log^{10} of average: $1 \times 10^{average} = 1 \times 10^{1.955\ 8} = 90.3 = 90$

Refer to American Public Health Association; American Water Works Association; and Water Environment Federation (1992) *Standard Methods for the Examination of Water and Wastewater*. 18th Ed., Washington, D.C., p. 9-12.

36. d. 1:1000

    1 mL into 99 mL = 1:100
    1 mL into 99 mL = 1:1000
    10 mL across filter = 1:10
    (1:100) × (1:100) × (1:10) = 1:100 000

Refer to American Public Health Association; American Water Works Association; and Water Environment Federation (1992) *Standard Methods for the Examination of Water and Wastewater*. 18th Ed., Washington, D.C., p. 9-35.

37. b. 63 mg/L

    Seeded BOD =

$$\dfrac{(\text{Initial Sample DO, mg/L} - \text{Final Sample DO, mg/L}) - [(\text{Initial Seed Control DO, mg/L} - \text{Final Seed Control DO, mg/L}) \times f]}{\left(\dfrac{\text{Sample Volume, mL}}{\text{Bottle Volume, mL}}\right)}$$

Where

$f = \dfrac{\text{Volume of Seed in Sample, mL}}{\text{Volume of Seed in Control, mL}}$

$$\text{Seeded BOD} = \frac{(8.0\ \text{mg/L} - 4.0\ \text{mg/L}) - [(8.1\ \text{mg/L} - 5.3\ \text{mg/L})(\frac{1.8\ \text{mL}}{6\ \text{mL}})]}{\frac{15\ \text{mL}}{300\ \text{mL}}}$$

$$= \frac{4.0\ \text{mg/L} - (2.8\ \text{mg/L} \times 0.3)}{0.05}$$

$$= \frac{4.0\ \text{mg/L} - 0.84\ \text{mg/L}}{0.05}$$

$$= \frac{3.16\ \text{mg/L}}{0.05}$$

$$= 63.2\ \text{mg/L (rounded to 63 mg/L)}$$

Refer to American Public Health Association; American Water Works Association; and Water Environment Federation (1992) *Standard Methods for the Examination of Water and Wastewater*. 18th Ed., Washington, D.C., p. 5-5.

38. a. 0.3

$$f = \frac{\text{Volume of Seed in Sample, mL}}{\text{Volume of Seed in Control, mL}}$$

$$= \frac{1.8\ \text{mL}}{6\ \text{mL}}$$

$$= 0.3$$

Refer to American Public Health Association; American Water Works Association; and Water Environment Federation (1992) *Standard Methods for the Examination of Water and Wastewater*. 18th Ed., Washington, D.C., p. 5-5.

39. b.   83 mg/L

$$\text{BOD} = \frac{(\text{Initial DO, mg/L} - \text{Final DO, mg/L})}{\dfrac{\text{Sample Volume, mL}}{\text{Bottle Volume, mL}}}$$

$$= \frac{(8.9 \text{ mg/L} - 6.4 \text{ mg/L})}{\left(\dfrac{9 \text{ mL}}{300 \text{ mL}}\right)}$$

$$= \frac{2.5 \text{ mg/L}}{0.03}$$

$$= 83.3 \text{ mg/L (rounded to 83 mg/L)}$$

Refer to American Public Health Association; American Water Works Association; and Water Environment Federation (1992) *Standard Methods for the Examination of Water and Wastewater.* 18th Ed., Washington, D.C., p. 5-5.

40. b.   400 mL

Dilution = (Initial Volume) (Initial Molarity) = (Final Volume) (Final Molarity)

Initial Volume (0.30 M) = (1000 mL) (0.12 M)

$$\text{Initial Volume} = \frac{(1000 \text{ mL})(0.12 \text{ M})}{0.30 \text{ M}}$$

$$= 400 \text{ mL}$$

Refer to Smith, R.-K. (1995) *Water and Wastewater Laboratory Techniques.* Water Environ. Fed., Alexandria, Va., p. 198.

41. c.   164 mg/L

$$\text{TSS} = \frac{(\text{Weight of Filter and Dry Residue, mg} - \text{Weight of Filter, mg})}{\text{Sample Volume, mL}} \times 1000 \frac{\text{mL}}{\text{L}}$$

$$= \frac{114.9 \text{ mg} - 110.8 \text{ mg}}{25 \text{ mL}} \times 1000 \frac{\text{mL}}{\text{L}}$$

$$= 164 \text{ mg/L}$$

Refer to American Public Health Association; American Water Works Association; and Water Environment Federation (1992) *Standard Methods for the Examination of Water and Wastewater.* 18th Ed., Washington, D.C., p. 2-54.

42. c. 83%

Sample with Spike − Sample = Calculated Recovery
2.66 mg/L − 0.83 mg/L = 1.83 mg/L
Percent Accuracy =
100% − [(Known Addition or Matrix Spike − Calculated Recovery) × 100%]
  = 100% − [(2.00 mg/L − 1.83 mg/L) × 100%]
  = 100% − 17%
  = 83%

American Public Health Association; American Water Works Association; and Water Environment Federation (1992) *Standard Methods for the Examination of Water and Wastewater.* 18th Ed., Washington, D.C., p. 1-10.

43. a. 175 g

MW of KCl = 75 g/mol
2.33 M = 2.33 mol/L

$$\text{Grams Needed} = (1\,\cancel{L})\left(75\,\frac{g}{\cancel{mol}}\right)\left(2.33\,\frac{\cancel{mol}}{\cancel{L}}\right)$$
$$= 174.75 \text{ g (rounded to 175 g)}$$

Refer to Smith, R.-K. (1995) *Water and Wastewater Laboratory Techniques.* Water Environ. Fed., Alexandria, Va., p. 197.

SAMPLING

Questions

1. Preservative is added to the sample container before the sample so that

    a. You will not forget to add the preservative later
    b. All sample portions are preserved as soon as they are collected
    c. There is less danger of coming into contact with the preservative or the sample
    d. It will not be necessary to carry caustic or acid around the plant or in the field

2. Filterable residue samples are subject to the following preservation, sample container, and hold time requirements:

    a. Preserve with H_2SO_4 to pH < 2 in plastic with a maximum hold time of 7 days
    b. Cool to 4 °C in glass with a maximum hold time of 7 days
    c. Cool to 4 °C in plastic with a maximum hold time of 7 days
    d. Cool to 4 °C in glass or plastic with a maximum hold time of 7 days

3. Sample preservation is intended to

    a. Retard biological action
    b. Increase volatility of constituents
    c. Accelerate hydrolysis of chemical compounds and complexes
    d. Remove trace metals from the sample

4. Failure to adhere to sampling procedures may result in

    a. Defensible analytical data

    b. Plant upset
    c. Invalid samples
    d. Representative samples

5. The primary objective of a sampling plan is to collect

    a. Representative samples that exhibit average properties of the entire waste stream while maintaining sample integrity so there is no significant change in sample chemistry before analysis
    b. Sample quantities that are practical for sampling and laboratory personnel to handle and that provide sufficient sample volume to perform most analyses
    c. Sample portions that allow measurements of the chemical properties of the waste that are neither accurate nor precise
    d. The entire waste stream

6. Sampling points in the wastewater plant should be located

    a. At the surface in turbulent areas specified in the regulatory permit
    b. In the center of the channel, at the bottom
    c. At a point where the flow is well mixed
    d. At a calm location where settling can occur

7. A sample is in your custody if

    a. It is within your sight or in your physical possession
    b. You have secured the sample in a laboratory refrigerator
    c. You have released the sample to an overnight courier
    d. The sample has been prepared for analysis

8. Analytical data must be defensible in court in the event of litigation. To ensure that your results are defensible, it is necessary to

    a. Restrict sample access to the minimum number of people possible
    b. Be able to trace sample possession and handling from the time of collection through analysis and final disposal
    c. Keep samples in locked tamperproof storage containers at all times
    d. Place a new custody seal on the sample container each time it is opened

9. The ability to trace sample possession and handling from the time of collection through analysis and final disposal is known as

    a. Tracking
    b. Sample identification
    c. Chain of custody
    d. Sample logging

10. Which of the following is an advantage of automatic sampling?

    a. The risk of human error may be increased
    b. Sampling events are missed because of bad weather conditions
    c. Sample aliquots can be collected at almost any preprogrammed rate of frequency
    d. Labor costs are increased

11. Cold storage of a grab sample is not necessary if BOD analysis is started

    a. Immediately, or the sample must be cooled to 4 °C
    b. Within 2 hours of sample collection
    c. Within 6 hours of sample collection
    d. Within 24 hours of sample collection

12. What is the most important information on a sample label?

    a. Date, origin, initials, and time
    b. Time, date, origin, and analysis
    c. Origin, time, date, and volume
    d. Initials, analysis, date, and volume

13. How can you collect a homogeneous sample of a wastewater stream?

    a. Take a sample where there is adequate mixing of the wastewater stream
    b. Do not remix the sample during compositing
    c. Do not preserve the sample
    d. Take a sample near the surface of the waste stream

14. What is the maximum time that a refrigerated sample may be held before BOD analysis?

    a. 6 hours
    b. 24 hours
    c. 48 hours
    d. 28 days

15. Why are flow-proportional composite samples collected?

    a. The flow and waste characteristics are continually changing
    b. The flow is continually changing
    c. The waste characteristics are continually changing
    d. The flow and waste characteristics never change

16. Composite samples can be used for determining

    a. pH
    b. Oil and grease
    c. Fecal coliform
    d. BOD

17. Preservation methods are limited to pH control, chemical addition, refrigeration, filtration, freezing, and

    a. Storage in amber opaque bottles
    b. Autoclaving
    c. Digestion
    d. Incubation

18. Which property need not be measured in the field because it is not subject to rapid changes?

    a. pH
    b. Dissolved gases
    c. Alkalinity
    d. Temperature

19. Samples collected for total phosphorus analysis are subject to which of the following preservation, sample container, and hold time requirements?

    a. Preserve with H_2SO_4 to pH < 2 in glass or plastic with a maximum hold time of 7 days
    b. Preserve with H_2SO_4 to pH < 2 in glass or plastic with a maximum hold time of 14 days

c. Preserve with H_2SO_4 to pH < 2 in glass or plastic with a maximum hold time of 28 days

d. Preserve with NaOH to pH > 12 in glass or plastic with a maximum hold time of 28 days

20. Samples collected for determination of ammonia nitrogen and nitrate–nitrite levels are subject to which of the following preservation, sample container, and hold time requirements?

a. Preserve with H_2SO_4 to pH < 2 in glass or plastic with a maximum hold time of 7 days

b. Preserve with H_2SO_4 to pH < 2 in glass or plastic with a maximum hold time of 14 days

c. Preserve with H_2SO_4 to pH < 2 in glass or plastic with a maximum hold time of 28 days

d. Preserve with NaOH to pH > 12 in glass or plastic with a maximum hold time of 28 days

21. A sample that is collected at a particular instant in time and reflects the source material conditions is a

a. Grab sample
b. Composite sample
c. Flow-proportional sample
d. Fixed-in-the-field sample

22. The most common preservation method is

   a. Acidification
   b. Eutrophication
   c. Cooling to 4 °C
   d. Dechlorination

Answers

1. b. All sample portions are preserved as soon as they are collected

   Refer to American Public Health Association; American Water Works Association; and Water Environment Federation (1992) *Standard Methods for the Examination of Water and Wastewater*. 18th Ed., Washington, D.C., p. 1-23.

2. d. Cool to 4 °C in glass or plastic with a maximum hold time of 7 days

   Refer to American Public Health Association; American Water Works Association; and Water Environment Federation (1992) *Standard Methods for the Examination of Water and Wastewater*. 18th Ed., Washington, D.C., p. 1-22.

3. a. Retard biological action

   Refer to American Public Health Association; American Water Works Association; and Water Environment Federation (1992) *Standard Methods for the Examination of Water and Wastewater*. 18th Ed., Washington, D.C., p. 1-23.

4. c. Invalid samples

   Refer to American Public Health Association; American Water Works Association; and Water Environment Federation (1992) *Standard Methods for the Examination of Water and Wastewater*. 18th Ed., Washington, D.C., p. 1-18.

5. a. Representative samples that exhibit average properties of the entire waste stream while maintaining sample integrity so there is no significant change in sample chemistry before analysis

Refer to American Public Health Association; American Water Works Association; and Water Environment Federation (1992) *Standard Methods for the Examination of Water and Wastewater.* 18th Ed., Washington, D.C., p. 1-18.

6. c. At a point where the flow is well mixed

Refer Water Environment Federation (1996) *Operation of Municipal Wastewater Treatment Plants.* 5th Ed., Manual of Practice No. 11, Alexandria, Va., p. 496.

7. a. It is within your sight or in your physical possession

Refer to American Public Health Association; American Water Works Association; and Water Environment Federation (1992) *Standard Methods for the Examination of Water and Wastewater.* 18th Ed., Washington, D.C., p. 1-20.

8. b. Be able to trace sample possession and handling from the time of collection through analysis and final disposal

Refer to American Public Health Association; American Water Works Association; and Water Environment Federation (1992) *Standard Methods for the Examination of Water and Wastewater.* 18th Ed., Washington, D.C., p. 1-20.

9. c. Chain of custody

Refer to American Public Health Association; American Water Works Association; and Water Environment Federation (1992) *Standard Methods for the Examination of Water and Wastewater.* 18th Ed., Washington, D.C., p. 1-20.

10. c. Sample aliquots can be collected at almost any preprogrammed rate of frequency

Refer to American Public Health Association; American Water Works Association; and Water Environment Federation (1992) *Standard Methods for the Examination of Water and Wastewater.* 18th Ed., Washington, D.C., p. 1-21.

11. b. Within 2 hours of sample collection

Refer to American Public Health Association; American Water Works Association; and Water Environment Federation (1992) *Standard Methods for the Examination of Water and Wastewater.* 18th Ed., Washington, D.C., p. 5-3.

12. a. Date, origin, initials, and time

Refer to American Public Health Association; American Water Works Association; and Water Environment Federation (1992) *Standard Methods for the Examination of Water and Wastewater.* 18th Ed., Washington, D.C., p. 1-20.

13. a. Take a sample where there is adequate mixing of the wastewater stream
Sampling where the stream mixes well and remixing before pouring the sample ensure a representative and homogeneous sample for analysis.

Refer to American Public Health Association; American Water Works Association; and Water Environment Federation (1992) *Standard Methods for the Examination of Water and Wastewater.* 18th Ed., Washington, D.C., page 1-19.

14. c. 48 hours

Refer to American Public Health Association; American Water Works Association; and Water Environment Federation (1992) *Standard Methods for the Examination of Water and Wastewater*. 18th Ed., Washington, D.C., p. 1-22.

15. a. The flow and waste characteristics are continually changing

Refer to California State University (1992) *Operation of Wastewater Treatment Plants*. 4th Ed., Sacramento, Calif., p. 479.

16. d. BOD

Refer to American Public Health Association; American Water Works Association; and Water Environment Federation (1992) *Standard Methods for the Examination of Water and Wastewater*. 18th Ed., Washington, D.C., p. 1-22.

17. a. Storage in amber opaque bottles

Refer to American Public Health Association; American Water Works Association; and Water Environment Federation (1992) *Standard Methods for the Examination of Water and Wastewater*. 18th Ed., Washington, D.C., p. 1-23.

18. c. Alkalinity

Refer to American Public Health Association; American Water Works Association; and Water Environment Federation (1992) *Standard Methods for the Examination of Water and Wastewater*. 18th Ed., Washington, D.C., p. 1-22.

19. c. Preserve with H_2SO_4 to pH < 2 in glass or plastic with a maximum hold time of 28 days

Refer to U.S. Environmental Protection Agency (1983) *Methods for Chemical Analysis of Water and Wastewater.* EPA-600/4-79-020 (revised March 1983), Environ. Monit. Support Lab., Cincinnati, Ohio, p. xviii.

20. c. Preserve with H_2SO_4 to pH < 2 in glass or plastic with a maximum hold time of 28 days

Refer to U.S. Environmental Protection Agency (1983) *Methods for Chemical Analysis of Water and Wastewater.* EPA-600/4-79-020, (revised March 1983), Environ. Monit. Support Lab., Cincinnati, Ohio, p. xvii.

21. a. Grab sample

Refer to Water Environment Federation (1996) *Operation of Municipal Wastewater Treatment Plants.* 5th Ed., Manual of Practice No. 11, Alexandria, Va., p. 491.

22. c. Cooling to 4 °C

Refer to American Public Health Association; American Water Works Association; and Water Environment Federation (1992) *Standard Methods for the Examination of Water and Wastewater.* 18th Ed., Washington, D.C., p. 1-22.

COMPLIANCE

Note that compliance is often a local issue. Please check with your local authority for regulations in your area. Some testing bodies add specific regulatory questions to examinations.

Questions

1. What is the relationship between the *Federal Register* and the CFR?

    a. The *Federal Register* announces changes to the CFR
    b. The *Federal Register* announces all of the court decisions based on the CFR
    c. The CFR publishes functions of Congress and the *Federal Register* addresses information related to the U.S. EPA
    d. There is no relationship between the two

2. Which part of 40 CFR concerns wastewater analysis?

    a. 136
    b. 141
    c. 258
    d. 264

3. A new method of analysis may be used provided that which of the following agencies can be convinced by parallel data that the new method is equivalent to the approved method?

    a. Standard Methods Board of Testing Procedures
    b. U.S. Bureau of Standards
    c. ASTM Committee for Testing Procedures
    d. U.S. EPA

4. Analysis performed as a result of a monitoring requirement imposed by a municipality, the state, or the federal government in support of an NPDES permit is known as

   a. CFR
   b. Process control testing
   c. Regulatory reporting
   d. *Federal Register*

Answers

1. a. The *Federal Register* announces changes to the CFR

    Refer to Smith, R.-K. (1999) *Lectures on Wastewater Analysis and Interpretation.* Genium Publishing, Schenectady, N.Y., p. 33.

2. a. 136

    Refer to Smith, R.-K. (1999) *Lectures on Wastewater Analysis and Interpretation.* Genium Publishing, Schenectady, N.Y., p. 5; and Smith, R.-K. (1999) *Handbook of Environmental Analysis.* 4th Ed., Genium Publishing, Schenectady, N.Y., p. 4.

3. d. U.S. EPA

    Refer to Smith, R.-K. (1999) *Lectures on Wastewater Analysis and Interpretation.* Genium Publishing, Schenectady, N.Y., p. 32; and Smith, R.-K. (1999) *Handbook of Environmental Analysis.* 4th Ed., Genium Publishing, Schenectady, N.Y., p. 58.

4. c. Regulatory reporting

    Refer to Water Environment Federation (1996) *Operation of Municipal Wastewater Treatment Plants.* 5th Ed., Manual of Practice No. 11, Alexandria, Va., p. 489.

CONDUCTIVITY

Questions

1. The U.S. EPA Conductance method is

    a. Method 120.1
    b. Method 150.1
    e. Method 160.2
    d. Method 413.1

2. Reporting units for conductivity are

    a. milligrams per liter at 25 °C
    b. micrograms per liter at 25 °C
    c. milligrams per kilogram at 25 °C
    d. millisiemens per meter at 25 °C

3. When performing a conductivity measurement, you should determine the temperature of the samples ± 0.5 °C. If the temperature of the sample is not 25 °C,

    a. Proceed with the analysis and adjust results for the temperature difference according to instructions in the method
    b. Collect a new sample and repeat temperature measurement
    c. Apply ice or heat on a hotplate to adjust sample temperature to 25 °C
    d. Proceed with the analysis and ignore any temperature variations

4. Preservation and hold time recommendations for samples collected for conductivity analysis are as follows

   a. Collect in glass or plastic and cool to 4 °C with a maximum hold time of 48 hours
   b. Collect in glass or plastic and cool to 4 °C with a maximum hold time of 28 days
   c. Collect in glass only and cool to 4 °C with a maximum hold time of 48 hours
   d. Collect in glass only and cool to 4 °C with a maximum hold time of 28 days

5. Conductivity measurements are performed at or corrected to a temperature of

   a. 4 °C
   b. 20 °C
   c. 25 °C
   d. 25 °F

6. The measurement of the ability of solutions to carry a current is called

   a. Voltage
   b. Conductivity
   c. Electricity
   d. Electrolytes

7. Conductivity measurements should be reported at what temperature?

   a. 20 ± 0.1 °C
   b. 20 ± 1 °C

c. 25 ± 0.1 °C

d. 25 ± 1 °C

8. A specific conductivity test will give you an indication of

   a. Total dissolved solids
   b. Soluble BOD
   c. Total organic carbon
   d. Insoluble heavy metals

Answers

1. a. Method 120.1

    Refer to U.S. Environmental Protection Agency (1983) *Methods for Chemical Analysis of Water and Wastewater.* EPA-600/4-79-020 (revised March 1983), Environ. Monit. Support Lab., Cincinnati, Ohio, p. 120.1-1.

2. d. millisiemens per meter at 25 °C

    Refer to U.S. Environmental Protection Agency (1983) *Methods for Chemical Analysis of Water and Wastewater.* EPA-600/4-79-020 (revised March 1983), Environ. Monit. Support Lab., Cincinnati, Ohio, p. 120.1-2.

3. a. Proceed with the analysis and adjust results for the temperature difference according to instructions in the method

    Refer to U.S. Environmental Protection Agency (1983) *Methods for Chemical Analysis of Water and Wastewater.* EPA-600/4-79-020 (revised March 1983), Environ. Monit. Support Lab., Cincinnati, Ohio, p. 120.1-2.

4. b. Collect in glass or plastic and cool to 4 °C with a maximum hold time of 28 days

    Refer to U.S. Environmental Protection Agency (1983) *Methods for Chemical Analysis of Water and Wastewater.* EPA-600/4-79-020 (revised March 1983), Environ. Monit. Support Lab., Cincinnati, Ohio, p. xvi.

5. c. 25 °C

Refer to American Public Health Association; American Water Works Association; and Water Environment Federation (1992) *Standard Methods for the Examination of Water and Wastewater*. 18th Ed., Washington, D.C., p. 2-46.

6. b. Conductivity

Refer to American Public Health Association; American Water Works Association; and Water Environment Federation (1992) *Standard Methods for the Examination of Water and Wastewater*. 18th Ed., Washington, D.C., p. 2-43.

7. d. 25 ± 0.1 °C

Refer to American Public Health Association; American Water Works Association; and Water Environment Federation (1992) *Standard Methods for the Examination of Water and Wastewater*. 18th Ed., Washington, D.C., p. 2-46.

8. a. Total dissolved solids

Refer to Water Environment Federation (1996) *Operation of Municipal Wastewater Treatment Plants*. 5th Ed., Manual of Practice No. 11, Alexandria, Va., p. 480.

QUALITY ASSURANCE/QUALITY CONTROL

Questions

1. Physical and chemical measurements used in laboratories should be selected by which of the following criterion?

    a. Should measure required constituents in the presence of routine interferences with sufficient precision and accuracy to meet data needs
    b. Should use equipment and skills not ordinarily available in the average laboratory
    a. Should be sufficiently tested by the laboratory manager
    d. Should be approved by outside laboratories

2. Responsibility for the results obtained from using a nonstandard procedure rests with the

    a. Analyst
    b. Laboratory supervisor and U.S. EPA coordinator
    c. U.S. EPA coordinator and plant superintendent
    d. U.S. EPA coordinator only

3. In a statistical formula, precision is estimated by

    a. Standard deviation or the range
    b. Arithmetic mean and the range
    c. Arithmetic mean and the percent recovery
    d. Percent recovery and the range

4. Percent recovery is expressed as

    a. parts per million

b. milligrams per liter
c. percent
d. parts per billion

5. A control chart must indicate the conditions under which it was developed. These do not include

   a. Date
   b. Method used
   c. Required reagents
   d. Laboratory name

6. Which condition below would NOT indicate an out-of-control situation?

   a. Seven successive data points alternating above and below the central line
   b. Seven successive data points just below the central line
   c. One data point beyond the upper control limit
   d. Seven successive data points alternating above and below the lower control limit

7. What should be the first response when an analysis is deemed out of control?

   a. Continue the analysis until another run is out of control
   b. Stop the analysis and identify and resolve the problem
   c. Recalculate the control limits so that the analysis will fall within the new control limits
   d. Stop the analysis and recalculate new control limits from the beginning

8. To verify the control chart, which of the following applies?

   a. If the distribution is proper, approximately 50% of the initializing data should fall within the interval $P \pm Sp$
   b. If the distribution is proper, approximately 68% of the initializing data should fall within the interval $P \pm Sp$
   c. If the distribution is proper, 95% of the initializing data should fall within the interval $P \pm Sp$
   d. If the distribution is proper, 100% of the initializing data should fall within the interval $P \pm Sp$

9. When should a new standard curve be established?

   a. Immediately after an analysis goes out of control
   b. Just before using up the stock reagents
   c. When making a change in equipment, method, or reagents
   d. After analyzing one set of samples

10. With each batch of analyses, which test should be run?

    a. Blank on water and reagents, standard recovery, and spike recovery
    b. Standard recovery and duplicate analysis
    c. Duplicate analysis and blank on water and reagents
    d. Blank on water and reagents, duplicate analysis, and standard recovery

11. Accuracy is

    a. A series of measurements in which one or more of the results vary greatly from the other values
    b. The percent probability that a measurement result will lie within the confidence interval

    c.   A measure of the degree of agreement among replicate analyses of a sample, typically expressed as the standard deviation

    d.   The combination of bias and precision of an analytical procedure that reflects the closeness of a measured value to a true value

12. When calculating control limits, data should be rejected if

    a.   They are all within the old control limits
    b.   They pass the outliers test
    c.   A known error has occurred
    d.   The result is suspicious

13. Quality assurance is NOT

    a.   A definitive plan for laboratory operation that specifies the measures used to produce data of known precision and bias
    b.   A set of measures within a sample analysis methodology to ensure that the process is in control
    c.   A set of operating principles that is strictly followed during sample collection and analysis
    d.   Data of known and defensible quality

14. To calculate the geometric mean, measurements must be converted to

    a.   Semilogs
    b.   Logarithms
    c.   Negative integers
    d.   Whole numbers

15. The standard deviation is

    a. The square root of the mean
    b. The square root of the variance
    c. The square root of the geometric mean
    d. The square root of the range

16. A good quality control program includes at least seven elements: certification of operator competence, recovery of known additions, analysis of externally supplied standards, analysis of reagent blanks, calibration with standards, analysis of duplicates, and

    a. An efficient sample log-in system
    b. Maintenance of control charts
    c. A comprehensive preventive maintenance program
    d. A reagent lot number logbook

17. Values found above the highest standard should not be reported unless

    a. An initial demonstration of greater linear range has been made, all instrument parameters have been changed, and the value is fewer than 1.5 times the highest standard
    b. An initial demonstration of greater linear range has been made, no instrument parameters have been changed, and the value is fewer than 1.5 times the highest standard
    c. The value is more than 1.5 times the highest standard
    d. An initial demonstration of greater linear range has been made and the value is fewer than 1.25 times the highest standard

18. Precision in analytical determination is a measure of

    a. Bias
    b. Accuracy
    c. Recovery
    d. Repeatability

19. The measure of a consistent deviation of measured values from the true value caused by systemic errors in a procedure is called

    a. Bias
    b. Accuracy
    c. Distribution
    d. Precision

20. A measure of bias and precision of an analytical procedure that reflects the closeness of a measured value to the true value is called

    a. Precision
    b. Accuracy
    c. Percent recovery
    d. Variance

21. A standard used to determine the state of instrument accuracy between periodic calibrations is called a

    a. Standard blank
    b. Laboratory standard
    c. Check standard
    d. Reference standard

22. To determine whether matrix effects influenced an analysis, the analyst should

   a. Run surrogate standards
   b. Run replicate samples
   c. Spike blanks
   d. Determine the recovery of known additions

23. At a minimum, how many calibration standards should be used when analysis is initiated?

   a. One
   b. Two
   c. Three
   d. Four

24. When using known additions for the assessment of analytical quality, the addition of known material should be

   a. Performed every 20 samples
   b. One and ten times the ambient level
   c. Of a pure solution
   d. Above the demonstrated linear range of the analysis

25. Quality control analysis of externally supplied standards is used to check

   a. Precision
   b. Bias
   c. Accuracy
   d. Method detection limit

26. On a control chart, how many standard deviations are the warning limit from the average percent recovery?

    a. 1
    b. 2
    c. 3
    d. 4

27. When a measurement exceeds the control limits, you should

    a. Repeat the analysis immediately
    b. Stop the analysis and look for problems
    c. Continue analysis and look for trends
    d. Recalibrate with new standards

28. On a quality control chart, what percentage of values should fall between the warning limits?

    a. 68%
    b. 75%
    c. 95%
    d. 99%

29. A good quality control program should include

    a. Duplicate analyses, blanks, and control standards
    b. Spiked samples, replicate analyses, and standard operating procedures
    c. Analysis of the same sample by different analysts, blanks, and duplicate analyses
    d. Replicate analyses, spiked samples, and analysis of the same sample by different analysts

30. Which of the following is the correct definition of "significant figures"?

    a. The number of digits to the right of the decimal point
    b. The total number of digits in a reported result
    c. All digits known to be true and one last digit in doubt
    e. The number of digits to the left of the decimal point

31. If TSS is reportable to two significant figures, how do you report the calculated value 109.3?

    a. 109
    b. 109.3
    c. 110
    d. 109.30

32. When a series of numbers is added, the sum should be rounded off to

    a. The same number of decimal places as the figure with the least number of decimal places
    b. The same number of decimal places as the figure with the highest number of decimal places
    c. Two decimal places regardless of the number of decimal places in the numbers being added
    d. Three decimal places regardless of the number of decimal places in the numbers

33. The following series of numbers is added: (23.46 + 2.666 + 5.1) = 31.226. The result should be rounded off and reported as

    a. 31
    b. 31.2

c. 31.226
d. 31.23

34. A BOD is calculated as 23.50 and should be rounded to two significant figures. The result should be reported as

   a. 23
   b. 23.0
   c. 24
   d. 24.0

35. Round the result of multiplication or division calculations to the number of significant figures present

   a. In the factor with the most significant figures
   b. In the factor with the least significant figures
   c. In the total number of decimal places in all of the factors
   d. In the smallest number

36. Another expression of the mean is the

   a. Standard deviation
   b. Coefficient of variation
   c. Average
   d. Median

37. The expression $[(x - \text{avg. } x)^2/(n \times 1)]^{1/2}$ indicates the estimate of the

   a. Mean
   b. Coefficient of variation

c. Variance
d. Standard deviation

38. On a means quality control chart, the control limit is how many standard deviations from the mean?

   a. 1
   b. 2
   c. 3
   d. 4

39. What is an aliquot of analyte-free water that is carried through the entire testing process called?

   a. Reagent blank
   b. Distilled water
   c. Matrix spike
   d. Standard

40. Graphic representations of quality control measurements performed on a regular basis are

   a. Percent recovery
   b. RPD
   c. Control charts
   d. Semilogarithmic graphs

Answers

1. a. Should measure required constituents in the presence of routine interferences with sufficient precision and accuracy to meet data needs

   Refer to U.S. Environmental Protection Agency (1979) *Handbook for Analytical Quality Control in Water and Wastewater Laboratories.* EPA-600/4-79-019, Environ. Monit. Support Lab., Cincinnati, Ohio, p. 1-3.

2. a. Analyst

   Refer to U.S. Environmental Protection Agency (1979) *Handbook for Analytical Quality Control in Water and Wastewater Laboratories.* EPA-600/4-79-019, Environ. Monit. Support Lab., Cincinnati, Ohio, p. 1-4.

3. a. Standard deviation or the range

   Refer to U.S. Environmental Protection Agency (1979) *Handbook for Analytical Quality Control in Water and Wastewater Laboratories.* EPA-600/4-79-019, Environ. Monit. Support Lab., Cincinnati, Ohio, p. 6-1.

4. c. percent

   Refer to U.S. Environmental Protection Agency (1979) *Handbook for Analytical Quality Control in Water and Wastewater Laboratories.* EPA-600/4-79-019, Environ. Monit. Support Lab., Cincinnati, Ohio, p. 2-2.

5. c. Required reagents
   Must include date, method used, laboratory name, parameter, and any other information unique to the initializing data, such as range of concentration and identification of analyst(s).

Refer to U.S. Environmental Protection Agency (1979) *Handbook for Analytical Quality Control in Water and Wastewater Laboratories.* EPA-600/4-79-019, Environ. Monit. Support Lab., Cincinnati, Ohio, p. 6-5.

6. a. Seven successive data points alternating above and below the central line
Any point beyond the control limits or any seven successive points on the same side of the centerline indicate an out-of-control situation.

Refer to U.S. Environmental Protection Agency (1979) *Handbook for Analytical Quality Control in Water and Wastewater Laboratories.* EPA-600/4-79-019, Environ. Monit. Support Lab., Cincinnati, Ohio, p. 6-5.

7. b. Stop the analysis and identify and resolve the problem
Afterwards, the frequency of quality control checks should increase for the next few runs.

Refer to U.S. Environmental Protection Agency (1979) *Handbook for Analytical Quality Control in Water and Wastewater Laboratories.* EPA-600/4-79-019, Environ. Monit. Support Lab., Cincinnati, Ohio, p. 6-5.

8. b. If the distribution is proper, approximately 68% of the initializing data should fall within the interval $P \pm Sp$

Refer to U.S. Environmental Protection Agency (1979) *Handbook for Analytical Quality Control in Water and Wastewater Laboratories.* EPA-600/4-79-019, Environ. Monit. Support Lab., Cincinnati, Ohio, p. 6-5.

9. c. When making a change in equipment, method, or reagents

   Refer to U.S. Environmental Protection Agency (1979) *Handbook for Analytical Quality Control in Water and Wastewater Laboratories.* EPA-600/4-79-019, Environ. Monit. Support Lab., Cincinnati, Ohio, p. 6-9.

10. a. Blank on water and reagents, standard recovery, and spike recovery

    Refer to U.S. Environmental Protection Agency (1979) *Handbook for Analytical Quality Control in Water and Wastewater Laboratories.* EPA-600/4-79-019, Environ. Monit. Support Lab., Cincinnati, Ohio, p. 6-11.

11. d. The combination of bias and precision of an analytical procedure that reflects the closeness of a measured value to a true value

    Refer to American Public Health Association; American Water Works Association; and Water Environment Federation (1992) *Standard Methods for the Examination of Water and Wastewater.* 18th Ed., Washington, D.C., p. 1-3.

12. c. A known error has occurred

    Refer to American Public Health Association; American Water Works Association; and Water Environment Federation (1992) *Standard Methods for the Examination of Water and Wastewater.* 18th Ed., Washington, D.C., p. 1-2.

13. b. A set of measures within a sample analysis methodology to ensure that the process is in control

    Refer to American Public Health Association; American Water Works Association; and Water Environment Federation (1992) *Standard Methods for the Examination of Water and Wastewater.* 18th Ed., Washington, D.C., p. 1-3.

14. b. Logarithms

Refer to U.S. Environmental Protection Agency (1979) *Handbook for Analytical Quality Control in Water and Wastewater Laboratories.* EPA-600/4-79-019, Environ. Monit. Support Lab., Cincinnati, Ohio, p. 7-4.

15. b. The square root of the variance

Refer to U.S. Environmental Protection Agency (1979) *Handbook for Analytical Quality Control in Water and Wastewater Laboratories.* EPA-600/4-79-019, Environ. Monit. Support Lab., Cincinnati, Ohio, p. 7-5.

16. b. Maintenance of control charts
The rest are elements of the broader quality assurance program.

Refer to American Public Health Association; American Water Works Association; and Water Environment Federation (1992) *Standard Methods for the Examination of Water and Wastewater.* 18th Ed., Washington, D.C., p. 1-4.

17. b. An initial demonstration of greater linear range has been made, no instrument parameters have been changed, and the value is fewer than 1.5 times the highest standard

Refer to American Public Health Association; American Water Works Association; and Water Environment Federation (1992) *Standard Methods for the Examination of Water and Wastewater.* 18th Ed., Washington, D.C., p. 1-5 and 6.

18. d. Repeatability

Refer to American Public Health Association; American Water Works Association; and Water Environment Federation (1992) *Standard Methods for the Examination of Water and Wastewater*. 18th Ed., Washington, D.C., p. 1-9.

19. a. Bias

Refer to American Public Health Association; American Water Works Association; and Water Environment Federation (1992) *Standard Methods for the Examination of Water and Wastewater*. 18th Ed., Washington, D.C., p. 1-9.

20. b. Accuracy

Refer to American Public Health Association; American Water Works Association; and Water Environment Federation (1992) *Standard Methods for the Examination of Water and Wastewater*. 18th Ed., Washington, D.C., p. 1-3.

21. c. Check standard

Refer to American Public Health Association; American Water Works Association; and Water Environment Federation (1992) *Standard Methods for the Examination of Water and Wastewater*. 18th Ed., Washington, D.C., p. 1-3.

22. d. Determine the recovery of known additions

Refer to American Public Health Association; American Water Works Association; and Water Environment Federation (1992) *Standard Methods for the Examination of Water and Wastewater*. 18th Ed., Washington, D.C., p. 1-4.

23. c. Three

Refer to American Public Health Association; American Water Works Association; and Water Environment Federation (1992) *Standard Methods for the Examination of Water and Wastewater*. 18th Ed., Washington, D.C., p. 1-5.

24. b. One and ten times the ambient level

Refer to American Public Health Association; American Water Works Association; and Water Environment Federation (1992) *Standard Methods for the Examination of Water and Wastewater*. 18th Ed., Washington, D.C., p. 1-5.

25. c. Accuracy

Refer to American Public Health Association; American Water Works Association; and Water Environment Federation (1992) *Standard Methods for the Examination of Water and Wastewater*. 18th Ed., Washington, D.C., p. 1-5.

26. b. 2

Refer to American Public Health Association; American Water Works Association; and Water Environment Federation (1992) *Standard Methods for the Examination of Water and Wastewater*. 18th Ed., Washington, D.C., p. 1-5.

27. a. Repeat the analysis immediately

Refer to American Public Health Association; American Water Works Association; and Water Environment Federation (1992) *Standard Methods for the Examination of Water and Wastewater*. 18th Ed., Washington, D.C., p. 1-7.

28. c. 95%

Refer to American Public Health Association; American Water Works Association; and Water Environment Federation (1992) *Standard Methods for the Examination of Water and Wastewater*. 18th Ed., Washington, D.C., p. 1-5.

29. a. Duplicate analyses, blanks, and control standards

Refer to American Public Health Association; American Water Works Association; and Water Environment Federation (1992) *Standard Methods for the Examination of Water and Wastewater*. 18th Ed., Washington, D.C., p. 1-4.

30. c. All digits known to be true and one last digit in doubt

Refer to American Public Health Association; American Water Works Association; and Water Environment Federation (1992) *Standard Methods for the Examination of Water and Wastewater*. 18th Ed., Washington, D.C., p. 1-17.

31. c. 110

There are four significant figures in 109.30 and 109.3 and three in 109.

Refer to American Public Health Association; American Water Works Association; and Water Environment Federation (1992) *Standard Methods for the Examination of Water and Wastewater*. 18th Ed., Washington, D.C., p. 1-17.

32. a. The same number of decimal places as the figure with the least number of decimal places

Refer to American Public Health Association; American Water Works Association; and Water Environment Federation (1992) *Standard Methods for the Examination of Water and Wastewater.* 18th Ed., Washington, D.C., p. 1-17.

33. b. 31.2

Refer to American Public Health Association; American Water Works Association; and Water Environment Federation (1992) *Standard Methods for the Examination of Water and Wastewater.* 18th Ed., Washington, D.C., p. 1-17.

34. c. 24

Figures ending in the number 5 followed only by zeros are rounded to the nearest even number.

Refer to American Public Health Association; American Water Works Association; and Water Environment Federation (1992) *Standard Methods for the Examination of Water and Wastewater.* 18th Ed., Washington, D.C., p. 1-17.

35. b. In the factor with the least significant figures

Refer to American Public Health Association; American Water Works Association; and Water Environment Federation (1992) *Standard Methods for the Examination of Water and Wastewater.* 18th Ed., Washington, D.C., p. 1-17.

36. c. Average

Refer to American Public Health Association; American Water Works Association; and Water Environment Federation (1992) *Standard Methods for the Examination of Water and Wastewater.* 18th Ed., Washington, D.C., p. 1-5.

37. d. Standard deviation

Refer to American Public Health Association; American Water Works Association; and Water Environment Federation (1992) *Standard Methods for the Examination of Water and Wastewater.* 18th Ed., Washington, D.C., p. 1-1.

38. c. 3

Refer to American Public Health Association; American Water Works Association; and Water Environment Federation (1992) *Standard Methods for the Examination of Water and Wastewater.* 18th Ed., Washington, D.C., p. 1-5.

39. a. Reagent blank

Refer to American Public Health Association; American Water Works Association; and Water Environment Federation (1992) *Standard Methods for the Examination of Water and Wastewater.* 18th Ed., Washington, D.C., p. 1-5.

40. c. Control charts

Refer to American Public Health Association; American Water Works Association; and Water Environment Federation (1992) *Standard Methods for the Examination of Water and Wastewater.* 18th Ed., Washington, D.C., inside front cover.

NITROGEN

Questions

1. Total Kjeldahl nitrogen is the sum of

    a. Ammonia, organic nitrogen, and nitrate nitrogen
    b. Ammonia, nitrite, and nitrate nitrogen
    c. Azide, nitrite, and organic nitrogen
    d. Ammonia and organic nitrogen

2. The Kjeldahl method is a method for determining

    a. pH concentration
    b. Nitrogen
    c. Chlorine
    d. H_2S

3. The most oxidized form of nitrogen that may be found in water and wastewater is

    a. Organic nitrogen
    b. Ammonia
    c. Nitrate
    d. Nitrite

4. The cadmium-reduction method is recommended for determining what concentration range of nitrate nitrogen?

    a. Less than 0.1 ppm
    b. 0.01 to 1.0 ppm
    c. 0.5 to 20 ppm
    d. 1 to 10 ppm

5. What is the interference source for the ammonia-specific ion electrode?

   a. Total organic carbon
   b. Alkalinity
   c. B_2O_3
   d. Amines

6. In the preliminary distillation step for Kjeldahl ammonia, the sample is buffered to a pH of

   a. Less than 4
   b. 7.0
   c. 9.5
   d. Greater than 10

7. The intermediate oxidation stage of nitrogen between ammonia and nitrate nitrogen is

   a. TKN
   b. Nitrite nitrogen
   c. Organic nitrogen
   d. Free nitrogen

8. For the ion-specific electrode method of ammonia analysis, how much sample must be used?

   a. 10 mL
   b. 50 mL
   c. 100 mL
   d. 500 mL

9. For the phenate method of ammonia analysis, at what wavelength should the spectrophotometer be set?

    a. 515 nm
    b. 530 nm
    c. 615 nm
    d. 630 nm

10. The cadmium-reduction technique is used when analyzing

    a. Ammonia nitrogen
    b. Nitrate nitrogen
    c. Nitrite nitrogen
    d. Nitrite and nitrate nitrogen

Answers

1. d. Ammonia and organic nitrogen

    Refer to American Public Health Association; American Water Works Association; and Water Environment Federation (1992) *Standard Methods for the Examination of Water and Wastewater.* 18th Ed., Washington, D.C., p. 4-75.

2. b. Nitrogen

    Refer to California Water Environment Association (1993) *Laboratory Analysts Study Manual.* Oakland, Calif., p. 6-4; and Smith, R.-K. (1999) *Lectures on Wastewater Analysis and Interpretation.* Genium Publishing, Schenectady, N.Y., p. 161.

3. c. Nitrate

    Refer to California Water Environment Association (1993) *Laboratory Analysts Study Manual.* Oakland, Calif., p. 7-7; and Smith, R.-K. (1999) *Lectures on Wastewater Analysis and Interpretation.* Genium Publishing, Schenectady, N.Y., p. 160.

4. b. 0.01 to 1.0 ppm

    Refer to American Public Health Association; American Water Works Association; and Water Environment Federation (1992) *Standard Methods for the Examination of Water and Wastewater.* 18th Ed., Washington, D.C., p. 4-89.

5. d. Amines

Refer to American Public Health Association; American Water Works Association; and Water Environment Federation (1992) *Standard Methods for the Examination of Water and Wastewater.* 18th Ed., Washington, D.C., p. 4-82.

6. c. 9.5

Refer to American Public Health Association; American Water Works Association; and Water Environment Federation (1992) *Standard Methods for the Examination of Water and Wastewater.* 18th Ed., Washington, D.C., p. 4-77.

7. b. Nitrite nitrogen

Refer to Smith, R.-K. (1999) *Lectures on Wastewater Analysis and Interpretation.* Genium Publishing, Schenectady, N.Y., p. 159.

8. c. 100 mL

Refer to American Public Health Association; American Water Works Association; and Water Environment Federation (1992) *Standard Methods for the Examination of Water and Wastewater.* 18th Ed., Washington, D.C., p. 4-82.

9. d. 630 nm

Refer to American Public Health Association; American Water Works Association; and Water Environment Federation (1992) *Standard Methods for the Examination of Water and Wastewater.* 18th Ed., Washington, D.C., p. 4-80.

10. b. Nitrate nitrogen

Refer to American Public Health Association; American Water Works Association; and Water Environment Federation (1992) *Standard Methods for the Examination of Water and Wastewater*. 18th Ed., Washington, D.C., p. 4-89.

OIL/GREASE

Questions

1. Proper sampling techniques for an oil and grease sample include

    a. Composite sample collected in a glass container
    b. Manually collected grab sample in a glass container
    c. Manually collected grab sample in glass or plastic container
    d. Refrigeration and preservation with 1:1 HNO_3

2. *Standard Methods* defines "oil and grease" as

    a. All petroleum-based products
    b. Any material recovered as a substance soluble in the solvent
    c. All animal and vegetable sources of fatty matter
    d. All light oils, heavy oils, volatile oils, and tars

3. The correct sample preservation and hold time for oil and grease samples are

    a. H_2SO_4 or HCl addition to < 2.0 pH with a hold time of 14 days
    b. H_2SO_4 or HCl addition to < 2.0 pH with a hold time of 28 days
    c. NaOH addition to > 10 pH with a hold time of 14 days
    d. NaOH addition to > 10 pH with a hold time of 28 days

4. In the Method 5520 B. (Partition–Gravimetric Method), solvents are distilled from flasks in a water bath maintained at

    a. 80 °C for trichlorotrifluoroethane and 85 °C for *n*-hexane
    b. 70 °C for trichlorotrifluoroethane and 85 °C for *n*-hexane

c. 70 °C for trichlorotrifluoroethane and 75 °C for *n*-hexane
d. 80 °C for trichlorotrifluoroethane and 75 °C for *n*-hexane

5. U.S. EPA recommends which of the following sample collection criterion for oil and grease analyses?

   a. Collection of 1-L composite samples in polyethylene bottles
   b. Collection of 1-L grab samples in glass bottles
   c. Preservation with NaOH
   d. No refrigeration

6. Which of the following is the proper collection technique for samples collected for oil and grease analysis?

   a. Collect composite samples using an automatic sampler and glass containers
   b. Collect a grab sample and store in a glass container
   c. Collect a grab sample and store in a plastic or glass container
   d. Refrigerate the sample and preserve with nitric acid to pH < 2

Answers

1. b. Manually collected grab sample in a glass container

    Refer to American Public Health Association; American Water Works Association; and Water Environment Federation (1992) *Standard Methods for the Examination of Water and Wastewater.* 18th Ed., Washington, D.C., p. 5-251.

2. b. Any material recovered as a substance soluble in the solvent

    This may include non-oil substances such as extractable sulfur, organic dyes, and chlorophyll, which are subject to the extraction procedure.

    Refer to American Public Health Association; American Water Works Association; and Water Environment Federation (1992) *Standard Methods for the Examination of Water and Wastewater.* 18th Ed., Washington, D.C., p. 5-24.

3. b. H_2SO_4 or HCl addition to < 2.0 pH with a hold time of 28 days

    Refer U.S. Code of Federal Regulations (1999) *Guidelines Establishing Test Procedures for Analysis of Pollutants Under Clean Water Act.* 40 CFR, Part 136, Washington, D.C.; and U.S. Environmental Protection Agency (1983) *Methods for Chemical Analysis of Water and Wastewater.* EPA-600/4-79-020 (revised March 1983), Environ. Monit. Support Lab., Cincinnati, Ohio, pp. xviii and 413.1-1.

4. b. 70 °C for trichlorotrifluoroethane and 85 °C for *n*-hexane

    Refer to American Public Health Association; American Water Works Association; and Water Environment Federation (1992) *Standard Methods for the Examination of Water and Wastewater.* 18th Ed., Washington, D.C., p. 5-26.

5. b. Collection of 1-L grab samples in glass bottles

Composite sampling results in the loss of grease on collection equipment. The analysis of individual grab samples and a reported average result is a more accurate "composite" value.

Refer to U.S. Environmental Protection Agency (1983) *Methods for Chemical Analysis of Water and Wastewater.* EPA-600/4-79-020 (revised March 1983), Environ. Monit. Support Lab., Cincinnati, Ohio, p. 413.1-1.

6. b. Collect a grab sample and store in a glass container

Refer to American Public Health Association; American Water Works Association; and Water Environment Federation (1992) *Standard Methods for the Examination of Water and Wastewater.* 18th Ed., Washington, D.C., p. 1-22

PHOSPHORUS

Questions

1. What is one of the adverse effects of discharging phosphorus to a receiving stream?

    a. It stimulates nuisance quantities of algal growth
    b. It raises the turbidity to an unhealthy level
    c. It reduces photosynthesis
    d. It causes a decrease in net primary production

2. The term "total phosphorus" refers to

    a. All phosphorus in the sample, regardless of form
    b. Phosphorus not combined with organic compounds
    c. Orthophosphate
    d. Condensed phosphate

3. The reason for acid digestion of the sample for analysis of total phosphorus is

    a. To eliminate interference from chlorine compounds
    b. To eliminate organic phosphate compounds
    c. To release phosphorus as orthophosphate
    d. To release the nitrogen compounds

Answers

1. a. It stimulates nuisance quantities of algal growth

    Refer to American Public Health Association; American Water Works Association; and Water Environment Federation (1992) *Standard Methods for the Examination of Water and Wastewater.* 18th Ed., Washington, D.C., p. 4-108.

2. a. All phosphorus in the sample, regardless of form

    Refer to California State University (1992) *Operation of Wastewater Treatment Plants.* 4th Ed., Sacramento, Calif., p. 573.

3. c. To release phosphorus as orthophosphate

    Refer to American Public Health Association; American Water Works Association; and Water Environment Federation (1992) *Standard Methods for the Examination of Water and Wastewater.* 18th Ed., Washington, D.C., p. 4-108; and Smith, R.-K. (1999) *Lectures on Wastewater Analysis and Interpretation.* Genium Publishing, Schenectady, N.Y., p. 190.

INORGANICS

Questions

1. How frequently should dilute metal standards solutions be prepared?

    a. Every analytical run
    b. Once a month
    c. Bimonthly
    d. Every six months

2. If oxidizing agents are not removed before preserving a cyanide sample, the agent may oxidize most of the cyanide in the sample. An example of an oxidizing agent is

    a. NaOH
    b. Chlorine
    c. H_2SO_4
    d. HCl

3. A 1-L sample of process water known to contain chlorine is collected for total cyanide analysis. To properly preserve the sample, what is added to the sample to reduce oxidizing agents before preserving with NaOH?

    a. Ascorbic acid
    b. Sodium arsenite
    c. KOH
    d. HCl

4. Holding time for a properly preserved cyanide sample is

    a. 3 days
    b. 7 days
    c. 14 days
    d. 21 days

5. What will distill with cyanide during distillation and adversely affect the colorimetric procedure?

    a. Sulfide
    b. Nitrogen
    c. NaOH
    d. Phenols

6. "Total metals" includes all metals

    a. Left on the filter after filtration
    b. In the filtrate after filtration
    c. Dissolved and particulate
    d. That can be detected in a gas chromatograph

7. Which of the following is used as a fuel gas in atomic absorption spectrophotometry?

    a. N_2O
    b. Helium
    c. Argon
    d. Oxygen

8. To which of the following analytical techniques does the Lambert–Beers law apply?

    a. Ion-selective electrode analysis
    b. Atomic absorption spectrophotometry
    c. Mass spectrometry
    d. Amperometric titrations

9. What are the two common oxidation states of chromium?

    a. +1 and +6
    b. +2 and +4
    c. +3 and +6
    d. +4 and +5

10. A 100-mL volume blank is digested and diluted to a final volume of 100.0 mL and analyzed to give 0.080 mg/L of chromium. A 1.00-g sample of solid is digested and diluted to a final volume of 100.0 mL and, when tested, exhibits 6.00 mg/kg chromium. Is the chromium in the sample the result of laboratory contamination or is it real?

    a. Chromium is really in the sample
    b. Chromium in the sample is completely the result of laboratory contamination
    c. It is impossible to tell
    d. It is actually vanadium incorrectly identified as chromium

Answers

1. a. Every analytical run

   Dilute metal solutions are prone to plate out on container walls over long periods of storage. Thus, dilute metal standard solutions must be prepared fresh at the time of analysis.

   Refer to U.S. Environmental Protection Agency (1979) *Handbook for Analytical Quality Control in Water and Wastewater Laboratories*. EPA-600/4-79-019, Environ. Monit. Support Lab., Cincinnati, Ohio, p. 4-2.

2. b. Chlorine

   Refer to American Public Health Association; American Water Works Association; and Water Environment Federation (1992) *Standard Methods for the Examination of Water and Wastewater*. 18th Ed., Washington, D.C., p. 4-21.

3. a. Ascorbic acid

   Refer to U.S. Code of Federal Regulations (1999) *Guidelines Establishing Test Procedures for Analysis of Pollutants Under Clean Water Act*. 40 CFR, Part 136, Washington, D.C., Table II.

4. c. 14 days

   Refer to American Public Health Association; American Water Works Association; and Water Environment Federation (1992) *Standard Methods for the Examination of Water and Wastewater*. 18th Ed., Washington, D.C., p. 1-22; and U.S. Code of Federal Regulations (1999) *Guidelines Establishing Test*

Procedures for Analysis of Pollutants Under Clean Water Act. 40 CFR, Part 136.3, Washington, D.C., Table II.

5. a. Sulfide

Refer to American Public Health Association; American Water Works Association; and Water Environment Federation (1992) *Standard Methods for the Examination of Water and Wastewater.* 18th Ed., Washington, D.C., p. 4-21.

6. c. Dissolved and particulate

Refer to American Public Health Association; American Water Works Association; and Water Environment Federation (1992) *Standard Methods for the Examination of Water and Wastewater.* 18th Ed., Washington, D.C., p. 3-1.

7. a. N_2O

Refer to U.S. Environmental Protection Agency (1983) *Methods for Chemical Analysis of Water and Wastewater.* EPA-600/4-79-020 (revised March 1983), Environ. Monit. Support Lab., Cincinnati, Ohio, p. METALS-10.

8. b. Atomic absorption spectrophotometry

The Lambert–Beers law relates concentration of analyte in solution to absorbance of light by the solution, $A = abc$, where A is absorbance, a is the extinction coefficient, b is the path length, and c is the concentration. Linear calibrations in light absorbance analytical procedures are based on the Lambert–Beers law.

Refer to Smith, R.-K. (1999) *Handbook of Environmental Analysis.* 4th Ed., Genium Publishing, Schenectady, N.Y., p. 297; and Smith, R.-K. (1999) *Lectures on Wastewater Analysis and Interpretation.* Genium Publishing, Schenectady, N.Y., p. 53.

9. c. +3 and +6

Refer to Smith, R.-K. (1999) *Handbook of Environmental Analysis.* 4th Ed., Genium Publishing, Schenectady, N.Y., p. 253.

10. b. Chromium in the sample is completely the result of laboratory contamination

Blank: 100 mL sample = 100 mL analytical digestate. Sample: 1.00 g sample = 100 mL analytical digestate. Blank result ×100 is a comparable value. 0.080 mg/L × 100 = 8.0 mg/L, which is greater than sample result of 6.0 mg/kg. Thus, all chromium detected in the sample is the result of laboratory contamination.

Refer to Smith, R.-K. (1999) *Handbook of Environmental Analysis.* 4th Ed., Genium Publishing, Schenectady, N.Y., p. 253.

ORGANICS

Questions

1. In a gas chromatograph, the component peaks are identified on the basis of

    a. The solvent used for extraction
    b. Peak height
    c. Retention time
    d. Peak area

2. Which gas chromatography detector is used to detect halogen-containing compounds?

    a. ECD
    b. FID
    c. PID
    d. MAD

3. Which gas chromatography detector uses UV light to detect compounds?

    a. ECD
    b. FID
    c. PID
    d. MAD

4. Which gas chromatography detector uses a flame to detect organic compounds?

    a. ECD
    b. FID
    c. PID
    d. MAD

5. Which component is not a part of a gas chromatograph?

    a. Column
    b. Detector
    c. Oven
    d. Nebulizer

6. Organic analytes are determined by which of the following analytical techniques?

    a. Gas chromatography
    b. Inductively coupled plasma atomic emission spectrophotometry
    c. Amperometric titration
    d. Heating and gravimetric analysis

7. Which pair of compounds are aromatics?

    a. Perchloroethane and trichloroethane
    b. Toluene and ethyl benzene
    c. NaCl and Na_2SO_4
    d. Benzene and $CHCl_3$

8. What is the percent of hydrogen by weight in hexadecane?

    a. 15.04%
    b. 17.71%
    c. 34.00%
    d. 68.00%

Answers

1. c. Retention time

    Refer to American Public Health Association; American Water Works Association; and Water Environment Federation (1992) *Standard Methods for the Examination of Water and Wastewater.* 18th Ed., Washington, D.C., p. 6-94; and Smith, R.-K. (1999) *Lectures on Wastewater Analysis and Interpretation.* Genium Publishing, Schenectady, N.Y., p. 83.

2. a. ECD

    Refer to American Public Health Association; American Water Works Association; and Water Environment Federation (1992) *Standard Methods for the Examination of Water and Wastewater.* 18th Ed., Washington, D.C., p. 6-4; Smith, R.-K. (1999) *Lectures on Wastewater Analysis and Interpretation.* Genium Publishing, Schenectady, N.Y., p. 301; and Smith, R.-K. (1999) *Handbook of Environmental Analysis.* 4th Ed., Genium Publishing, Schenectady, N.Y., p. 304.

3. c. PID

    Refer to American Public Health Association; American Water Works Association; and Water Environment Federation (1992) *Standard Methods for the Examination of Water and Wastewater.* 18th Ed., Washington, D.C., p. 6-5; and Smith, R.-K. (1999) *Lectures on Wastewater Analysis and Interpretation.* Genium Publishing, Schenectady, N.Y., p. 81; and Smith, R.-K. (1999) *Handbook of Environmental Analysis.* 4th Ed., Genium Publishing, Schenectady, N.Y., p. 301.

4.   b.   FID

Refer to American Public Health Association; American Water Works Association; and Water Environment Federation (1992) *Standard Methods for the Examination of Water and Wastewater.* 18th Ed., Washington, D.C., p. 6-4; Smith, R.-K. (1999) *Lectures on Wastewater Analysis and Interpretation.* Genium Publishing, Schenectady, N.Y., p. 81; and Smith, R.-K. (1999) *Handbook of Environmental Analysis.* 4th Ed., Genium Publishing, Schenectady, N.Y., p. 301.

5.   d.   Nebulizer

Refer to American Public Health Association; American Water Works Association; and Water Environment Federation (1992) *Standard Methods for the Examination of Water and Wastewater.* 18th Ed., Washington, D.C., p. 6-3; Smith, R.-K. (1999) *Lectures on Wastewater Analysis and Interpretation.* Genium Publishing, Schenectady, N.Y., p. 79; and Smith, R.-K. (1999) *Handbook of Environmental Analysis.* 4th Ed., Genium Publishing, Schenectady, N.Y., p. 304.

6.   a.   Gas chromatography

Refer to Smith, R.-K. (1999) *Lectures on Wastewater Analysis and Interpretation.* Genium Publishing, Schenectady, N.Y., p. 75.

7.   b.   Toluene and ethyl benzene

The term aromatic is used to indicate that there is an aromatic (benzene) ring in the molecule. Answer b. is the only choice in which both chemicals contain aromatic rings.

Refer to American Public Health Association; American Water Works Association; and Water Environment Federation (1992) *Standard Methods for the Examination of Water and Wastewater.* 18th Ed., Washington, D.C., p. 6-41.

8.   a.   15.04%

MW of $C_{16}H_{34}$

MW of C = 12 × 16 = 192

MW of H = 34 × 1 = +34
                                226

$$\frac{34}{226} \times 100\% = 15.04\%$$

Refer to Smith, R.-K. (1995) *Water and Wastewater Laboratory Techniques.* Water Environ. Fed., Alexandria, Va., p. 191.

GLOSSARY

absolute error measure of accuracy that is calculated as the difference between an observed or measured value and a true value.

absorption (1) Taking up of matter in bulk by other matter, as in dissolving of a gas by a liquid. (2) Penetration of substances into the bulk of the solid or liquid. See also *adsorption*.

accuracy The absolute nearness to the truth. In physical measurements, it is the degree of agreement between the quantity measured and the actual quantity. It should not be confused with "precision," which denotes the reproducibility of the measurement.

acid (1) A substance that tends to lose a proton. (2) A substance that dissolves in water with the formation of hydrogen ions. (3) A substance containing hydrogen which may be replaced by metals to form salts.

acidity The quantitative capacity of aqueous solutions to neutralize a base; measured by titration with a standard solution of a base to a specified endpoint; usually expressed as milligrams of equivalent calcium carbonate per liter (mg/L $CaCO_3$); not to be confused with pH. Water does not have to have a low pH to have high acidity.

adsorption The adherence of a gas, liquid, or dissolved material to the surface of a solid or liquid. It should not be confused with *absorption*.

aerobic Requiring, or not destroyed by, the presence of free or dissolved oxygen in an aqueous environment.

agar A gelatinous substance extracted from a red algae, commonly used as a medium for laboratory cultivation of bacteria.

agglomeration Coalescence of dispersed suspended matter into larger flocs or particles.

aliquot The amount of sample used for analysis.

alkaline The condition of water, wastewater, or soil that contains a sufficient amount of alkali substances to raise the pH above 7.0.

alkalinity The capacity of water to neutralize acids; a property imparted by carbonates, bicarbonates, hydroxides, and occasionally borates, silicates, and phosphates. It is expressed in milligrams of equivalent calcium carbonate per liter (mg/L $CaCO_3$).

anaerobic (1) A condition in which free and dissolved oxygen is unavailable. (2) Requiring or not destroyed by the absence of air or free oxygen.

analysis (1) Separation of a compound into its constituent parts. (2) The breaking down of a complex substance into simpler substances (quantitative analysis is the determination of the proportions of the constituents).

analyte Compound determined by an analysis.

analytical balance Mechanical or electronic balance having a sensitivity of at least 0.1 mg.

anhydrous Dry or devoid of water.

arithmetic mean The sum of a set of observations divided by the number of observations.

ash The nonvolatile inorganic solids that remain after incineration.

assimilate To absorb and incorporate as part of the cell.

atomic absorption spectrophotometry A highly sensitive instrumental technique for measuring trace quantities of elements in water.

atomic weight (1) Weight of an element relative to carbon. (2) The weight given in the periodic table.

autotrophic organisms Organisms including nitrifying bacteria and algae that use carbon dioxide as a source of carbon for cell synthesis. They can consume dissolved nitrates and ammonium salts.

bacteria A group of universally distributed, rigid, essentially unicellular microscopic organisms lacking chlorophyll. They perform a variety of biological treatment processes, including biological oxidation, sludge digestion, nitrification, and denitrification.

balance Instrument used in the laboratory to obtain precise weights to within 0.1 mg.

basic Alkaline; solution containing hydroxyl ions (OH) and having a pH greater than 7.0.

Beer's law States that, as the concentration of a light-absorbing species increases, the absorption of light will increase proportionately.

bias Consistent deviation of measured values from the true value caused by systematic errors in the procedure.

bicarbonate Acid salt of carbonic acid containing the radical HCO_3; component of the total alkalinity.

bioaccumulation Uptake and retention of substances by an organism from its surrounding environment and food.

bioassay (1) An assay method using a change in biological activity as a qualitative or quantitative means of analyzing a material's response to biological treatment. (2) A method of determining the toxic effects of industrial wastes and other wastewaters by using viable organisms; exposure of fish to various levels of a chemical under controlled conditions to determine safe and toxic levels of that chemical.

biochemical oxygen demand (BOD) A measure of the quantity of oxygen used in the biochemical oxidation of organic matter in a specified time, at a specific temperature, and under specified conditions.

biomass The mass of biological material contained in a system.

biota All living organisms within a system.

blank Artificial sample designed to monitor the introduction of artifacts to the measurement process. For aqueous samples, reagent water is used as a blank matrix. A universal matrix does not exist for solid samples and, therefore, no matrix is routinely used. There are several types of blanks that monitor a variety of processes. A laboratory blank is taken through sample preparation and analysis only. It tests for contamination in sample preparation and analyses. A field blank is opened in the field and tests for contamination from the atmosphere and those activities listed under trip blank. A trip blank is shipped to and from the field with the sample containers. It is not opened in the field and, therefore, provides a test for contamination from sample preservation; site conditions; transport; and sample storage, preparation, and analysis. A wash blank is poured appropriately over or through sample collection devices and tests for the cleanliness of sampling equipment and those activities listed under field blank.

broth Solution used in cultivation of microorganisms.

buffer A substance that resists a change in pH.

buffer capacity curve Plot of alkalinity of acidity titration volumes versus pH.

buret A glass tube with fine gradations and a bottom stopcock used to measure and dispense fluids accurately.

carbonaceous biochemical oxygen demand (CBOD) A quantitative measure of the amount of dissolved oxygen required for the biological oxidation of carbon-containing compounds in a sample. See *BOD*.

Celsius The international name for the centigrade scale of temperature, on which the freezing point and boiling point of water are 0 °C and 100 °C, respectively, at a barometric pressure of 1.013×10^5 Pa (760 mm Hg).

chain of custody Document designed to trace the custody of a sample(s) from the point of origin to final disposition, with the intent of legally proving that custody remained intact and that tampering or substitutions were precluded.

chemical oxygen demand (COD) A quantitative measure of the amount of oxygen required for the chemical oxidation of carbonaceous (organic) material in wastewater using inorganic dichromate or permanganate salts as oxidants in a 2-hour test.

chlorine (Cl_2) An element ordinarily existing as a greenish-yellow gas about 2.5 times heavier than air. At atmospheric pressure and a temperature of -30.1 °F (-48 °C), the gas becomes an amber liquid about 1.5 times heavier than water. Its atomic weight is 35.457, and its molecular weight is 70.914.

chromatography The generic name of a group of separation processes that depend on the redistribution of the molecules of a mixture between a gas or liquid phase in contact with one or more bulk phases. The types of chromatography are adsorption, column, gas, gel, liquid, thin-layer, and paper chromatography.

ciliated protozoa Protozoans with cilia (hairlike appendages) that assist in movement; common in trickling filters and healthy activated sludge. Free-swimming ciliates are present in the bulk liquid; stalked ciliates are commonly attached to solids matter in the liquid.

coagulation The conversion of colloidal (< 0.001 mm) or dispersed (0.001 to 0.1 mm) particles into small visible coagulated particles (0.1 to 1 mm) by the addition of a coagulant, compressing the electrical double layer surrounding each suspended particle, decreasing the magnitude of repulsive electrostatic interactions between particles, and thereby destabilizing the particle. See also *flocculation*.

colorimetry Analytical procedure involving development of color in a sample and measurement of color intensity as an indicator of concentration.

comparability Degree of confidence with which one set of data can be compared to a related set of data.

completeness Measure of the amount of valid data obtained from a measurement system relative to the amount expected to be obtained under correct, normal conditions.

composite sample A combination of individual samples of water or wastewater taken at preselected intervals to minimize the effect of the variability of the individual sample. Individual samples may be of equal volume or may be proportional to the flow at time of sampling.

compost The product of the thermophilic biological oxidation of sludge or other materials.

compound A substance consisting of two or more independent elements that can only be separated by chemical reactions.

concentrate Opposite of dilute; to increase the proportional amount of a material in solution by removal of water.

concentration (1) The amount of a given substance dissolved in a discrete unit volume of solution or applied to a unit weight of solid. (2) The process of increasing the dissolved solids per unit volume of solution, usually by evaporation of the liquid. (3) The process of increasing the suspended solids per unit volume of sludge as by sedimentation or dewatering.

condensation The process by which a substance changes from the vapor state to the liquid or solid state. Water that falls as precipitation from the atmosphere has condensed from the vapor state to rain or snow. Dew and frost are also forms of condensation.

condenser Any device for reducing gases or vapors to liquid or solid form (e.g., glass laboratory apparatus or glass tube enveloped by a water jacket used for distillation and reflux procedures).

conductivity Ability to carry an electrical current; also known as specific conductance.

continuing calibration Periodic analysis of one or more initial calibration standards to verify that the relationship established in the initial calibration continues to be valid.

corrosion The gradual deterioration or destruction of a substance or material by chemical action, frequently induced by electrochemical processes. The action proceeds inward from the surface.

crucible Small porcelain vessel used for ignition and fusion at high temperature.

cylinder Graduated cylinder; glass container graduated in milliliters to contain solutions and that is used for measuring liquids.

denitrification The anaerobic biological reduction of nitrate nitrogen to nitrogen gas; also, removal of total nitrogen from a system. See also *nitrification*.

desiccant A substance capable of absorbing moisture; used as a drying agent.

desiccator Glass vessel or cabinet containing a desiccant material for drying and storage of materials in a moisture-free environment.

detection Act of measuring the presence and amount of component(s) present in a sample (techniques commonly used are titration, gravimetric, spectroscopy, and chromatography).

detritus (1) Decaying organic matter such as root hairs, stems, and leaves typically found on the bottom of a water body. (2) Grit or fragments of rock or minerals.

digestion (1) The biological decomposition of the organic matter in sludge, resulting in partial liquefaction, mineralization, and volume reduction. (2) The process carried out in a digester. (3) Preparatory procedure in which solids in a sample are chemically solubilized. (4) Chemical treatment to convert material under test to a desired oxidation state or molecular configuration.

dilution (1) Lowering the concentration of a solution by adding more solvent. (2) The engineered mixing of discharged water with receiving water to lessen its immediate aesthetic and/or biochemical effect.

distillation Analytical procedure in which a material is removed from a solution as a gas by heating or boiling and recovered in a more nearly pure state by condensation as a liquid.

electrolytes Substances that exist as ions in a water solution (e.g., NaCl exists as Na^+ and Cl^-).

electron Negatively charged component of an atom.

electron capture detector (ECD) Gas chromatography detector that measures a decrease in electrical signal rather than an increase in electrical current. A GC carrier gas, normally nitrogen or argon/methane, flows through the detector. A H^{-3} or Ni^{63} beta particle-emitting radioactive source ionizes the carrier gas molecules, producing electrons that are measured at steady current. When a sample molecule (especially one containing halogen atoms] passes through the detector, the current is reduced because of "capture" of electrons by the halogen-containing molecule. The loss of current is proportional to the amount of compound present. The ECD is virtually insensitive to hydrocarbons. It is especially valuable for the analysis of pesticides and polychlorinated biphenyls.

element An irreducible constituent or constituent that cannot be readily separated into its components.

emulsion A heterogeneous liquid mixture of two or more liquids not normally dissolved in one another, but held in suspension one in the other by forceful agitation or by emulsifiers that modify the surface tension of the droplets to prevent coalescence.

endothermic A process or reaction that takes place with absorption of heat.

endpoint Point at which titration is completed, marked by a sudden change in color or deflection in reaction rate during a titration.

equivalent weight Weight of a compound that contains 1 g atom of available hydrogen or its chemical equivalent. The equivalent weight is found by

$$EW = \frac{MW}{Z}$$

Where
EW = equivalent weight,
MW = molecular weight, and
Z = a positive integer, the value for which depends on the valence (number of replaceable electrons).

evaporation (1) The process by which water becomes a vapor. (2) The quantity of water that is evaporated; the rate is expressed in depth of water, measured as liquid water removed from a specified surface per unit of time, generally in inches or centimeters per day, month, or year. (3) The concentration of dissolved solids by driving off water through the application of heat.

exothermic A process or reaction that is accompanied by the evolution of heat.

extraction The process of dissolving and separating out particular constituents of a liquid by treatment with solvents specific for those constituents. Extraction may be liquid☐solid or liquid☐liquid.

facultative Having the ability to live under different conditions; for example, with or without free oxygen.

filterable Capable of passing through a filter.

filtrate The liquid that has passed through a filter.

flagellates Microorganisms that move by the whipping action of a taillike projection called a *flagella*.

flame-ionization detector (FID) Gas chromatography detector in which the column effluent gas is mixed with hydrogen and burned in air or oxygen. The ions and electrons produced in the flame produce an electric current proportional to the amount of material in the detector. The FID responds to nearly all organic compounds but does not respond to air and water, making it exceptionally suited to environmental analysis.

flocculation In water and wastewater treatment, the agglomeration of colloidal and finely divided suspended matter after coagulation by gentle stirring by either mechanical or hydraulic means. For biological wastewater treatment in which coagulation is not used, agglomeration may be accomplished biologically.

fluorescence detector Highly sensitive (liquid crystal) detector that operates by exciting the LC column effluent stream with short wavelength radiation (typically UV) and monitoring the longer wavelength radiation emitted because of the presence of a specific organic chemical group. The fluorescence detector is a specific and selective detector, especially useful for detecting polycyclic aromatic hydrocarbons in environmental samples.

food-to-microorganism (F:M) ratio In the activated-sludge process, the loading rate expressed as pounds of BOD_5 per pound of mixed liquor or mixed liquor volatile suspended solids per day (lb BOD_5/d/lb MLSS or MLVSS).

formazin Finely divided material used to prepare stable emulsions for use as standards in turbidity measurement.

gas chromatography A method of separating a mixture of compounds into its constituents so they can be identified. The sample is vaporized into a gas-filled column, fractionated by being swept over a solid adsorbent, selectively eluted, and identified.

gas chromatography-mass spectrometry (GC-MS) An analytical technique involving the use of both gas chromatography and mass spectrometry, the former to separate a complex mixture into its components and the latter to deduce the atomic and molecular weights of those components. It is particularly useful in identifying organic compounds.

geometric mean Used to report average bacteria concentrations; determined by summing the logs of individual test results and dividing by the number of testings and taking the antilog.

Gooch crucible A heat-resistant container fitted with a filter mat used for determination of suspended and total solids.

grab sample A sample taken at a given place and time. It may be representative of the flow. See also *composite sample*.

graphite furnace atomic absorption spectrophotometer that heats the sample within a graphite tube using an electrical current; also commonly called a flameless furnace or heated graphite atomizer.

gravimetric Pertaining to the measurement of the weight of samples or materials.

grease and oil In wastewater, a group of substances including fats, waxes, free fatty acids, calcium and magnesium soaps, mineral oils, and certain other nonfatty materials; water-insoluble organic compounds of plant and animal origins or industrial wastes that can be removed by natural flotation skimming.

growth medium Broth or agar substrate composed of known amounts of nutrients necessary to support a particular organism or group of organisms.

hazardous chemical Chemical for which there is statistically significant evidence, based on at least one study conducted in accordance with established scientific principle, that acute or chronic health effects may occur if employees are exposed to the chemical.

hazardous waste Any waste that is potentially damaging to human or environmental health because of toxicity, ignitability, corrosivity, chemical reactivity, or other reasons.

heavy metals Metals that can be precipitated by hydrogen sulfide in acid solution, for example, lead, silver, gold, mercury, bismuth, and copper.

heterotrophic Microorganisms using organic matter as a substrate for growth (e.g., *Escherichia coli*).

homogenous Uniform in composition.

hydrolysis Break up of a substance into other or lesser substances by reaction with water.

hygroscopic Materials that readily absorb moisture.

ignition Quantitative combustion of a specimen to determine volatile and ash concentrations.

Imhoff cone A cone-shaped graduated vessel used to measure the volume of settleable solids in various liquids of wastewater origin during various settling times.

in situ Latin for in place.

inductively coupled argon plasma (ICP) Instrument used for metals analysis. Because the temperature of the plasma is considerably higher (10 000 K) than the temperature of a flame atomic absorption spectrophotometer, it is especially useful for refractory metals. It is also capable of multielement analysis.

industrial waste Waste generated by manufacturing or industrial practices that is not a hazardous waste regulated under Subtitle C of the Resource Conservation and Recovery Act.

initial calibration Analysis of standards containing varying amounts of analyte to establish the ratio of measurement system response to analyte mass or concentration across the working range of the analytical technique.

inorganic All those combinations of elements that do not include organic carbon.

inorganic matter Mineral-type compounds that are generally nonvolatile, not combustible, and not biodegradable. Most inorganic-type compounds or reactions are ionic in nature; therefore, rapid reactions are characteristic.

internal standard Method for quantifying chromatographic data in which a known amount (concentration) of one or more standard compounds is added to a sample before analysis. The amounts of various sample components are measured by comparing the peak areas of the components to the areas of appropriate closely eluting "added" internal standards. Response factors for each sample component relative to an appropriate internal standard are required. These response factors are obtained by analyses of standard solutions containing the organic components and the internal standards.

iodometric Group of analytical procedures based on foundation and titration of free iodine with thiosulfate.

ion A charged atom, molecule, or radical that affects the transport of electricity through an electrolyte or, to a certain extent, through a gas. An atom or molecule that has lost or gained one or more electrons.

ionization The dissociation of molecules into negatively and positively charged ions.

Lambert's law States that the absorption of light traveling through a light-absorbing medium will increase proportionately to the distance it travels through the medium.

larva Early, immature growth stage in metamorphic development.

liquid chromatography Chromatographic separation technique in which the substance to be analyzed is dissolved in a solvent and, using the same or a different solvent, is eluted through a solid adsorbent exhibiting differential adsorption for components of the substance.

mass spectrum Plot of ion mass-to-charge ratio versus intensity. A fragmentation pattern results from the impact of a beam of electrons on a given molecule, which produces a family of particles whose mass distribution is characteristic of the parent molecule. Qualitative and quantitative information are provided by a mass spectrum.

matrix Physical characteristics or state of a sample, for example, water, soil, or sludge.

matrix interference Influence of the sample matrix or sample components on the ability to qualitatively identify and quantitatively measure compounds in environmental samples.

mean (1) The arithmetic average of a group of data. (2) The statistical average (50% point) determined by probability analysis.

median In a statistical array, the value having as many cases larger in value as cases smaller in value.

method detection limit The constituent concentration that, when processed through the complete method, produces a sample with a 99% probability that it is different from the blank.

mineral acidity Component of the total acidity in wastewater; contributed by strong inorganic acids and measured by titrating with a base to a pH 4.5 endpoint.

mixture Substance containing two or more ingredients that retain their own individual properties.

mode Most frequently occurring value in a set of observations.

molarity Molar solution containing 1 g/mol of solute per liter of solution.

molecules Collection of atoms bound together by electrostatic and electromagnetic forces.

monitoring (1) Routine observation, sampling, and testing of designated locations or parameters to determine the efficiency of treatment or compliance with standards or requirements. (2) The procedure or operation of locating and measuring radioactive contamination by means of survey instruments that can detect and measure, as dose rate, ionizing radiations.

muffle furnace Electric furnace for ignition of specimens at high temperatures in an oxygen-deficient atmosphere.

muriatic acid Hydrochloric acid (HCl).

National Pollutant Discharge Elimination System (NPDES) A permit that is the basis for the monthly monitoring reports required by most states in the United States.

nematode Member of the phylum (Nematoda) of elongated cylindrical worms parasitic in animals or plants or free-living in soil or water.

Nernst equation Relationship among the electrode potentials, temperatures, and concentrations of species.

neutral Neither acidic nor alkaline; having a pH of 7.0 at 25 °C.

nitrification The oxidation of ammonia nitrogen to nitrate nitrogen in wastewater by biological or chemical reactions. See also *denitrification*.

nitrification inhibitor Chemical added to dilution water to prevent oxygen uptake by nitrifying bacteria present in the sample.

nonfilterable Materials capable of being retained by a filter.

normal solution A solution that contains one equivalent weight of a substance per liter of solution.

organic Refers to volatile, combustible, and sometimes biodegradable chemical compounds containing carbon atoms (carbonaceous) bonded together with other elements. The principal groups of organic substances found in wastewater are proteins, carbohydrates, and fats and oils. See also *inorganic*.

orthophosphate (1) A salt that contains phosphorus as $(PO_4)^{3-}$. (2) A product of hydrolysis of condensed (polymeric) phosphates. (3) A nutrient required for plant and animal growth.

oven Drying oven; used in the laboratory for drying specimens at a selected, constant above ambient temperature.

oxidation (1) A chemical reaction in which the oxidation number (valence) of an element increases because of the loss of one or more electrons by that element. Oxidation of an element is accompanied by simultaneous reduction of the other reactant. See also *reduction*. (2) The conversion of organic materials to simpler, more stable forms with the release of energy. This may be accomplished by chemical or biological means. (3) The addition of oxygen to a compound.

oxidation-reduction potential (ORP) The potential required to transfer electrons from the oxidant to the reductant and used as a qualitative measure of the state of oxidation in wastewater treatment systems.

oxygen uptake rate The oxygen used during biochemical oxidation, typically expressed as mg O_2/L·h in the activated-sludge process.

parts per million (ppm) The number of weight or volume units of a minor constituent present with each 1 million units of a solution or mixture. The more specific term milligrams per liter (mg/L) is preferred.

percent error Measure of accuracy that is calculated as the absolute error relative to the true value, expressed as a percent.

percent recovery Measure of accuracy that is calculated as the measured value relative to the true value, expressed as a percent.

performance audit Quantitative evaluation of a measurement system that involves the analysis of standard reference samples or materials that are certified as to their chemical composition or physical characteristics.

permissible exposure Maximum allowable peak, if no other measurable permissible exposure limits occur. Permissible exposure limits are time-weighted averages unless otherwise noted.

petri dish A covered dish containing agar media used in the laboratory to cultivate bacteria.

pH A measure of the hydrogen-ion concentration in a solution, expressed as the logarithm (base ten) of the reciprocal of the hydrogen-ion activity in gram moles per liter (g/mol/L). On the pH scale (0 to 14), a value of 7 at 25 °C (77 °F) represents a neutral condition. Decreasing values indicate increasing hydrogen-ion concentration (acidity); increasing values indicate decreasing hydrogen-ion concentration (alkalinity).

phosphorus An essential chemical element and nutrient for all life forms. Occurs in orthophosphate, pyrophosphate, tripolyphosphate, and organic phosphate forms. Each of these forms and their sum (total phosphorus) is expressed as milligrams per liter (mg/L) elemental phosphorus.

photoionization detector (PID) Gas chromatography detector in which the GC column effluent gas stream is subjected to UV radiation energetic enough to ionize many organic compounds in the gas stream The ions produced create a current between two electrodes. This current is proportional to the amount of compound present. Because each organic molecule has a characteristic ionization potential (the energy required to remove an electron from a molecule), the PID can be tuned to detect certain classes of organic compounds.

photometric Group of analytical procedures based on measurement of transmittance or absorbance of light of known wavelength through a solution.

pipet A calibrated glass tube used to deliver prescribed volumes of liquids, typically less than 10 mL.

plankton Aggregate of passively floating, drifting, or weakly motile organisms in a body of water (the organisms are mostly microscopic).

polychlorinated biphenyls (PCBs) A class of aromatic organic compounds with two six-carbon unsaturated rings, with chlorine atoms substituted on each ring and more than two such chlorine atoms per molecule of PCB. They are typically stable, resist both chemical and biological degradation, and are toxic to many biological species.

potentiometric Group of analytical procedures based on measurement of changes in electrical potential of a solution.

precipitate (1) To condense and cause to fall as precipitation, as water vapor condenses and falls as rain. (2) The separation from solution as a precipitate. (3) The substance that is precipitated.

precision Agreement or responsibility of a set of replicate results among themselves, typically expressed in terms of the deviation of a set of results from the arithmetic mean (precision may be qualified in terms of possible sources of variability, reproducibility, and repeatability).

preservative Either a chemical or reagent added to a sample to prevent or slow decomposition or degradation of a target analyte or physical process (such as cooling) used for the same purpose (both physical and chemical preservation may be used in tandem to prevent sample deterioration).

qualitative Identification of elements or components forming a substance; characterization of a material without reference to measured quantities or proportional amounts of various components.

quality assurance A definitive plan for operation that specifies the measures used to produce products or data of known precision and bias.

quality control Set of measurements within an analysis methodology to ensure that the process is in control.

quantitative Having to do with quantity; determination of actual weight, concentration, or proportional amount of a specific material.

random error deviation in any step in an analytical procedure that can be treated by standard statistical techniques.

range A measure of the variability of a quantity; the difference between the largest and smallest values in the sequence of values of the quantity.

reagent A chemical added to a system to bring about a chemical reaction.

reduction The addition of electrons to a chemical entity decreasing its valence. See also *oxidation*.

reflux (1) Cyclic distillation and condensation of a solution in a closed-loop system. (2) To perform extended boiling without evaporation loss.

refractory metals Metallic compounds resistant to temperature, corrosion, and treatment (e.g., borides, carbon, nitrides, oxides and silicides). See also *inductively coupled argon plasma*.

relative standard Measure of precision that is calculated as the standard deviation(s) of a set of values, relative to their arithmetic mean ($\bar{x}$) expressed as a percent (also referred to as coefficient of variation).

repeatability Precision of repeated measurements made on the same sample in the same laboratory at different times.

reportable limit Minimum concentration that must be reported.

resistance Opposition to electric current characteristic of some materials.

respiration rate Amount of oxygen being used by microorganisms in an activated-sludge system (oxygen uptake rate) during a specific time period; results of this analysis are normally expressed as milligrams oxygen per liter per minute.

rosolic acid Used to suppress a variety of nonfecal coliform organisms that might normally grow at elevated temperatures.

sample Must be representative of the material being sampled, uncontaminated by the sampling technique or container, of adequate size for laboratory examination, and properly and completely identified.

Sedgwick–Rafter cell Microscopic slide having a 1-mL reservoir and cover for examination and counting of organisms.

seed Small amount of substance or solution used to introduce a bacterial population to a sample that does not have sufficient microorganisms present.

self-purification Natural process occurring in a stream or other body of water that results in the reduction of bacteria and satisfaction of BOD.

sensitivity Ability of a measurement system to detect and accurately quantify a parameter at a critical level within a specific sample matrix.

separatory funnel Globular glass container with stopped neck, stopcock, and drain tube used for extraction of solutions with a solvent.

Series 600 Approved methods for organic chemical analysis of municipal and industrial wastewater; an appendix to 40 CFR 136.

sludge (1) The accumulated solids separated from liquids during the treatment process that have not undergone a stabilization process. (2) The removed material resulting from chemical treatment, coagulation, flocculation, sedimentation, flotation, or biological oxidation of water or wastewater. (3) Any solid material containing large amounts of entrained water collected during water or wastewater treatment.

sludge volume index (SVI) The ratio of the volume (in milliliters) of sludge settled from a 1000-mL sample in 30 minutes to the concentration of mixed liquor (in milligrams per liter [mg/L]) multiplied by 1000.

slurry A thick watery mud or any substance resembling it, such as lime slurry.

soluble Capable of being dissolved in a fluid.

solution A liquid that contains dissolved solute.

solvent Liquid capable of dissolving or dispersing one or more substances.

specific gravity The ratio of the mass of a body to the mass of an equal volume of water at a specific temperature, typically 20 °C (68 °F).

spectrophotometer An instrument for measuring the amount of electromagnetic radiation absorbed by a sample as a function of wavelength.

standard (1) Substance or solution of exact known chemical composition and concentration. (2) The accepted basis for comparison.

standard deviation Square root of the variance of a set of values.

stoichiometric The ratio of chemical substances reacting in water that corresponds to their combined weights in the theoretical chemical reaction.

supernatant (1) The liquid remaining above a sediment or precipitate after sedimentation. (2) The most liquid stratum in a sludge digester.

surrogate standard Organic compound similar in chemical composition to analytes of interest but not normally found in environmental samples. A known amount of surrogate standard is spiked into samples before sample preparation. They provide information about precision and accuracy of measurements. The use of surrogate standards is limited to chromatography techniques.

tare Deduction of gross weight to allow for the weight of the container or wrapper, used in gravimetric testing.

time-weighted average Average airborne exposure by employees in any 8-hour work shift of a 40-hour work week that cannot be exceeded.

titration The determination of a constituent in a known volume of solution by the measured addition of a solution of known strength to completion of the reaction as signaled by observation of an endpoint.

titrimetric Group of analytical procedures based on a titrated value.

to contain (TC) pipet Type of pipet usually used when a high degree of measurement accuracy is not required; the fluid remaining in the tip after draining must be blown out to accurately deliver the full measured amount.

to deliver (TC) pipet Type of pipet calibrated to deliver the calibrated volume of the pipet with a small drop left in the tip; provides a high degree of measurement accuracy.

total metals Those metals determined to be in a sample after rigorous digestion by physical and chemical means.

total organic carbon (TOC) The amount of carbon bound in organic compounds in a sample. Because all organic compounds have carbon as the common element, total organic carbon measurements provide a fundamental means of assessing the degree of organic pollution.

toxic wastes Wastes that can cause an adverse response when they come in contact with a biological entity.

trip balance single or double pan balance used in the laboratory for coarse weight to 0.1 g of accuracy.

turbidimetric Group of analytical procedures based on measurement of turbidity change affected by addition of chemical reagents.

ultimate biochemical oxygen demand (BOD_u) (1) Commonly, the total quantity of oxygen required to completely satisfy the first-stage BOD. (2) More strictly, the quantity of oxygen required to completely satisfy both the first-stage and second-stage BOD.

variance Sum of the squares of the difference between the individual values of a set and the arithmetic mean of the set divided by one fewer than the number of values composing the set.

viscosity The molecular attractions within a fluid that make it resist a tendency to deform under applied forces.

volatile Capable of being evaporated at relatively low temperatures.

volumetric pipet Pipet calibrated to contain and deliver a specified volume; also referred to as a TD pipet.

Whipple disc An ocular micrometer used in conjunction with the microscope; a scale scan through the eyepiece used for measurements.

REFERENCES

American Public Health Association; American Water Works Association; and Water Environment Federation (1992) *Standard Methods for the Examination of Water and Wastewater.* 18th Ed., Washington, D.C.

California State University (1992) *Operation of Wastewater Treatment Plants.* 4th Ed., Sacramento, Calif.

California Water Environment Association (1993) *Laboratory Analysts Study Manual.* Oakland, Calif.

Orion Research, Inc. *Orion Ross® Sure-Flow Electrodes Instruction Manual.* 227355-001 Rev A, Beverly, Mass.

Smith, R.-K. (1995) *Water and Wastewater Laboratory Techniques.* Water Environ. Fed., Alexandria, Va.

Smith, R.-K. (1999) *Handbook of Environmental Analysis.* 4th Ed., Genium Publishing, Schenectady, N.Y.

Smith, R.-K. (1999) *Lectures on Wastewater Analysis and Interpretation.* Genium Publishing, Schenectady, N.Y.

U.S. Code of Federal Regulations (1999) *Guidelines Establishing Test Procedures for Analysis of Pollutants Under Clean Water Act.* 40 CFR, Part 136, Washington, D.C.

U.S. Code of Federal Regulations, *Standards for the Use and Disposal of Sewage Sludge.* 40 CFR, Parts 257, 403, and 503, Washington, D.C.

U.S. Environmental Protection Agency (1979) *Handbook for Analytical Quality Control in Water and Wastewater Laboratories.* EPA-600/4-79-019, Environ. Monit. Support Lab., Cincinnati, Ohio

U.S. Environmental Protection Agency (1983) *Methods for Chemical Analysis of Water and Wastewater.* EPA-600/4-79-020 (revised March 1983), Environ. Monit. Support Lab., Cincinnati, Ohio

U.S. Environmental Protection Agency (1982) *Handbook for Sampling and Sample Preservation of Water and Wastewater.* EPA-600/4-82-029, Environ. Monit. Support Lab., Cincinnati, Ohio

Water Environment Federation (1996) *Operation of Municipal Wastewater Treatment Plants*. 5th Ed., Manual of Practice No. 11, Alexandria, Va.

Water Environment Federation (2000) *Operations Training CD-ROM Series/Activated Sludge Process Control* [CD-ROM]. Alexandria, Va.

APPENDIX A TABLE OF ELEMENTS

Symbol	Element	Atomic number	Atomic weight	Symbol	Element	Atomic number	Atomic weight
Ac	Actinium	89	227.03	Er	Erbium	68	167.26
Al	Aluminum	13	26.982	Eu	Europium	63	151.96
Am	Americium	95	243.06	Fm	Fermium	100	257.10
Sb	Antimony	51	121.76	F	Flourine	9	18.998
Ar	Argon	18	39.948	Fr	Francium	87	223.02
As	Arsenic	33	74.922	Gd	Gadolinium	64	157.25
At	Astatine	85	209.99	Ga	Gallium	31	69.723
Ba	Barium	56	137.33	Ge	Germanium	32	72.61
Bk	Berkelium	97	247.07	Au	Gold	79	196.97
Be	Beryllium	4	9.01208	Hf	Hafnium	72	178.49
Bi	Bismuth	83	208.98	Hs	Hassium	108	265.13
Bh	Bohrium	107	262.12	He	Helium	2	4.0026
B	Boron	5	10.811	Ho	Holmium	67	164.93
Br	Bromine	35	79.904	H	Hydrogen	1	1.0079
Cd	Cadmium	48	112.41	In	Indium	49	114.82
Ca	Calcium	20	40.078	I	Iodine	53	126.90
Cf	Californium	98	251.08	Ir	Iridium	77	192.22
C	Carbon	6	12.011	Fe	Iron	26	55.845
Ce	Cerium	58	140.12	Kr	Krypton	36	83.80
Cs	Cesium	55	132.91	La	Lanthanum	57	138.91
Cl	Chlorine	17	35.453	Lr	Lawrencium	103	262.11
Cr	Chromium	24	51.996	Pb	Lead	82	207.20
Co	Cobalt	27	58.933	Li	Lithium	3	6.941
Cu	Copper	29	63.546	Lu	Lutetium	71	174.97
Cm	Curium	96	247	Mg	Magnesium	12	24.305
Db	Dubnium	105	262.11	Mn	Manganese	25	54.938
Dy	Dysprosium	66	162.50	Mt	Meitnerium	109	266
Es	Einsteinium	99	252.08	Md	Mendelevium	101	258.10

Symbol	Element	Atomic number	Atomic weight	Symbol	Element	Atomic number	Atomic weight
Hg	Mercury	80	200.59	Sm	Samarium	62	150.36
Mo	Molybdenum	42	95.94	Sc	Scandium	21	44.956
Nd	Neodymium	60	144.24	Sg	Seaborgium	106	263.12
Ne	Neon	10	20.180	Se	Selenium	34	78.96
Np	Neptunium	93	237.05	Si	Silicon	14	28.086
Ni	Nickel	28	58.693	Ag	Silver	47	107.87
Nb	Niobium	41	92.906	Na	Sodium	11	22.990
N	Nitrogen	7	14.007	Sr	Strontium	38	87.62
No	Nobelium	102	259.10	S	Sulfur	16	32.066
Os	Osmian	76	190.23	Ta	Tantalum	73	180.95
O	Oxygen	8	15.999	Tc	Technetium	43	97.907
Pd	Palladium	46	106.42	Te	Tellurium	52	127.60
P	Phosphorus	15	30.974	Tb	Terbium	65	158.93
Pt	Platinum	78	195.08	Tl	Thalium	81	204.38
Pu	Plutonium	94	244.06	Th	Thorium	90	232.04
Po	Polonium	84	208.98	Tm	Thulium	69	168.93
K	Potassium	19	39.098	Sn	Tin	50	118.71
Pr	Praseodymium	59	140.91	Ti	Titanium	22	47.87
Pm	Promethium	61	144.91	W	Tungsten	74	183.84
Pa	Protactinium	91	231.04	U	Uranium	92	238.03
Ra	Radium	88	226.03	V	Vanadium	23	50.942
Rn	Radon	86	222.02	Xi	Xenon	54	131.29
Re	Rhenium	75	186.21	Yb	Ytterbium	70	173.04
Rh	Rhodium	45	102.91	Y	Yttrium	39	88.906
Rb	Rubidium	37	85.468	Zn	Zinc	30	65.39
Ru	Ruthenium	44	101.07	Zr	Zirconium	40	91.224
Rf	Rutherfordium	104	261.11				

APPENDIX B APPROVED METHODS

Environmental Protection Agency §136.3

TABLE IA.—LIST OF APPROVED BIOLOGICAL METHODS

Parameter and units	Method[1]	EPA	Standard methods, 18th Ed.	ASTM	USGS
Bacteria:					
1. Coliform (fecal), number per 100 mL.	Most Probable Number (MPN), 5 tube 3 dilution, or Membrane filter (MF)[2], single step	p. 132[3] p. 124[3]	9221C E[4] 9222D[4]	....	B-0050-85[5]
2. Coliform (fecal) in presence of chlorine, number per 100 mL.	MPN, 5 tube, 3 dilution, or MF, single step[6]	p. 132[3] p. 124[3]	9221C E[4] 9222D[4]		
3. Coliform (total), number per 100 mL.	MPN, 5 tube, 3 dilution, or MF[2] single step or two step	p. 114[3] p. 108[3]	9221B[4] 9222B[4]	....	B-0025-85[5]
4. Coliform (total), in presence of chlorine, number per 100 mL.	MPN, 5 tube, 3 dilution, or MF[2] with enrichment	p. 114[3] p. 111[3]	9221B[4] 9222(B+B.5c)[4]		
5. Fecal streptococci, number per 100 mL.	MPN, 5 tube, 3 dilution MF[2], or Plate count	p. 139[3] p. 136[3] p. 143[3]	9230B[4] 9230C[4]	....	B-0055-85[5]
Aquatic Toxicity:					
6. Toxicity, acute, fresh water organisms, LC50, percent effluent.	Daphnia, Ceriodaphnia, Fathead Minnow, Rainbow Trout, Brook Trout, or Bannerfish Shiner mortality.	Sec. 9[7]		....	
7. Toxicity, acute, estuarine and marine organisms, LC50, percent effluent.	Mysid, Sheepshead Minnow, or Menidia spp. mortality	Sec. 9[7]		....	
8. Toxicity, chronic, fresh water organisms, NOEC or IC25, percent effluent.	Fathead minnow larval survival and growth Fathead minnow embryo-larval survival and teratogenicity Ceriodaphnia survival and reproduction Selenastrum growth	1000.0[8] 1001.0[8] 1002.0[8] 1003.0[8]			
9. Toxicity, chronic, estuarine and marine organisms, NOEC or IC25, percent effluent.	Sheepshead minnow larval survival and growth Sheepshead minnow embryo-larval survival and teratogenicity Menidia beryllina larval and growth Mysidopsis bahia survival, growth, and fecundity Arbacia punctulata fertilization Champia parvula reproduction	1004.0[9] 1005.0[9] 1006.0[9] 1007.0[9] 1008.0[9] 1009.0[9]			

Notes to Table IA:
[1] The method must be specified when results are reported.
[2] A 0.45 um membrane filter (MF) or other pore size certified by the manufacturer to fully retain organisms to be cultivated and to be free of extractables which could interfere with their growth.
[3] USEPA. 1978. Microbiological Methods for Monitoring the Environment, Water, and Wastes. Environmental Monitoring and Support Laboratory. U.S. Environmental Protection Agency. Cincinnati, Ohio. EPA/600/8-78/017.
[4] APHA. 1992. Standard Methods for the Examination of Water and Wastewater. American Public Health Association. 18th Edition. Amer. Publ. Hlth. Assoc., Washington, DC.
[5] USGS. 1989. U.S. Geological Survey Techniques of Water-Resources Investigations. Book 5, Laboratory Analysis, Chapter A4, Methods for Collection and Analysis of Aquatic Biological and Microbiological Samples. U.S. Geological Survey. U.S. Department of Interior, Reston, Virginia.
[6] Because the MF technique usually yields low and variable recovery from chlorinated wastewaters, the Most Probable Number method will be required to resolve any controversies.
[7] USEPA. 1993. Methods for Measuring the Acute Toxicity of Effluents to Freshwater and Marine Organisms. Fourth Edition. Environmental Monitoring Systems Laboratory, U.S. Environmental Protection Agency, Cincinnati, Ohio. August 1993, EPA/600/4-90/027F.

§ 136.3

[a] USEPA. 1994. Short-term Methods for Estimating the Chronic Toxicity of Effluents and Receiving Waters to Freshwater Organisms. Third Edition. Environmental Monitoring Systems Laboratory, U.S. Environmental Protection Agency USEPA. 1994, Cincinnati, Ohio (July 1994, EPA/600/4-91/002).
[9] Short-term Methods for Estimating the Chronic Toxicity of Effluents and Receiving Waters to Marine and Estuarine Organisms. Second Edition. Environmental Monitoring Systems Laboratory, U.S. Environmental Protection Agency, Cincinnati, Ohio (July 1994, EPA/600/4-91/003). These methods do not apply to marine waters of the Pacific Ocean.

TABLE IB.—LIST OF APPROVED INORGANIC TEST PROCEDURES

Parameter, units and method	EPA [1,35]	STD methods 18th ed.	ASTM	USGS [2]	Other
1. Acidity, as $CaCO_3$, mg/L:					
Electrometric endpoint or phenolphthalein endpoint	305.1	2310 B(4a)	D1067-92		
2. Alkalinity, as $CaCO_3$, mg/L:					
Electrometric or Colorimetric titration to pH 4.5, manual or automated.	310.1 310.2	2320 B	D1067-92	I-1030-85 I-2030-85	973.43.[3]
3. Aluminum—Total,[4] mg/L; Digestion[4] followed by:					
AA direct aspiration [36]	202.1	3111 D			
AA furnace	202.2	3113 B			
ICP/AES [36]	[5]200.7	3120 B	D4190-82(88)	I-3051-85	
Inductively Coupled Plasma/Atomic Emission Spectrometry (ICP/AES) [36]					
Direct Current Plasma (DCP) [36]		3500–Al D			Note 34.
Colorimetric (Eriochrome cyanine R)					
4. Ammonia (as N), mg/L:					
Manual, distillation (at pH 9.5),[6] followed by	350.2	4500–NH_3B			
Nesslerization	350.2	4500–NH_3C	D1426-93(A)	I-3520-85	973.49.[3]
Titration	350.2	4500–NH_3E	D1426-93(B)		973.49.[3]
Electrode	350.3	4500–NH_3F or G			
Automated phenate, or	350.1	4500–NH_3H		I-4523-85	Note 7.
Automated electrode					
5. Antimony–Total,[4] mg/L; Digestion[4] followed by:					
AA direct aspiration [36]	204.1	3111 B			
AA furnace	204.2	3113 B			
ICP/AES [36]	[5]200.7	3120 B			
6. Arsenic–Total,[4] mg/L:					
Digestion[4] followed by	206.5				
AA gaseous hydride	206.3	3114 B 4.d	D2972-93(B)	I-3062-85	
AA furnace	206.2	3113 B	D2972-93(C)		
ICP/AES,[36] or	[5]200.7	3120 B			
Colorimetric (SDDC)	206.4	3500–As C	D2972-93(A)		
7. Barium—Total,[4] mg/L; Digestion[4] followed by:					
AA direct aspiration [36]	208.1	3111 D			
AA furnace	208.2	3113 B	D4382-91	I-3060-85	
ICP/AES [36]	[5]200.7	3120 B		I-3084-85	
DCP [36]					Note 34.
8. Beryllium—Total,[4] mg/L; Digestion[4] followed by:					
AA direct aspiration	210.1	3111 D	D3645-93(88)(A)	I-3095-85	
AA furnace	210.2	3113 B	D3645-93(88)(B)		
ICP/AES	[5]200.7	3120 B			
DCP, or			D4190-82(88)		Note 34.

Environmental Protection Agency § 136.3

Parameter					
9. Biochemical oxygen demand (BOD₅), mg/L:					
Colorimetric (aluminon)					
Dissolved Oxygen Depletion	405.1	5210 B			
10. Boron [37]—Total, mg/L:					
Colorimetric (curcumin)	212.3	4500-B B			Note 34
ICP/AES, or	[5]200.7	3120 B	D4190-82(88)		
DCP					
11. Bromide, mg/L:					
Titrimetric	320.1		D1246-82(88)(C)	I-1125-85	p. S44.[10]
12. Cadmium—Total,[4] mg/L; Digestion [4] followed by:					
AA direct aspiration [36]	213.1	3111 B or C	D3557-90(A or B)	I-3135-85 or I-3136-85	974.27,[3] p. 37.[9]
AA furnace	213.2	3113 B	D3557-90(D)		
ICP/AES [36]	[5]200.7	3120 B	D4190-82(88)	I-1472-85	Note 34.
DCP [36]			D3557-90(C)		
Voltametry,[11] or					
Colorimetric (Dithizone)		3500-Cd D			
13. Calcium—Total,[4] mg/L; Digestion [4] followed by:					
AA direct aspiration	215.1	3111 B	D511-93(B)	I-3152-85	
ICP/AES	[5]200.7	3120 B			
DCP, or					
Titrimetric (EDTA)	215.2	3500-Ca D	D511-93(A)		Note 34.
14. Carbonaceous biochemical oxygen demand (CBOD₅), mg/L [12]:		5210 B			
15. Chemical oxygen demand (COD), mg/L; Titrimetric, or	410.1	5220 C	D1252-88(A)	I-3560-85	973.46,[3] p. 17.[9]
	410.2			I-3562-85	
	410.3				
Spectrophotometric, manual or automated	410.4	5220 D	D1252-88(B)	I-3561-85	Notes 13 or 14.
16. Chloride, mg/L:					
Titrimetric (silver nitrate) or		4500-Cl⁻ B	D512-89(B)	I-1183-85	973.51.[3]
(Mercuric nitrate)	325.3	4500-Cl⁻ C	D512-89(A)	I-1184-85	
Colorimetric, manual or				I-1187-85	
Automated (Ferricyanide)	325.1 or 325.2	4500-Cl⁻ E		I-2187-85	
17. Chlorine—Total residual, mg/L; Titrimetric:					
Amperometric direct	330.1	4500-Cl D	D1253-86(92)		
Iodometric direct	330.3	4500-Cl B			
Back titration ether end-point [15] or	330.2	4500-Cl C			
DPD-FAS	330.4	4500-Cl F			
Spectrophotometric, DPD	330.5	4500-Cl G			
Or Electrode					Note 16.
18. Chromium VI dissolved, mg/L; 0.45 micron filtration followed by:					
AA chelation-extraction or	218.4	3111 C	D1687-92(A)	I-1232-85	
Colorimetric (Diphenylcarbazide)		3500-Cr D		I-1230-85	
19. Chromium—Total,[4] mg/L; Digestion [1] followed by:					
AA direct aspiration [36]	218.1	3111 B	D1687-92(B)	I-3236-85	974.27.[3]
AA chelation-extraction	218.3	3111 C			
AA furnace	218.2	3113 B	D1687-92(C)		

285

§ 136.3 40 CFR Ch. I (7–1–01 Edition)

TABLE IB.—LIST OF APPROVED INORGANIC TEST PROCEDURES—Continued

Parameter, units and method	EPA [1,35]	STD methods 18th ed.	ASTM	USGS [2]	Other
ICP/AES [36]	[5]200.7	3120 B			
DCP,[36] or					Note 34.
Colorimetric (Diphenylcarbazide)		3500-Cr D			
20. Cobalt—Total,[4] mg/L; Digestion [4] followed by:					
AA direct aspiration	219.1	3111 B or C	D3558-90(A or B)	I-3239-85	p. 37.[9]
AA furnace	219.2	3113 B	D3558-90(C)		
ICP/AES	[5]200.7	3120 B	D4190-82(88)		Note 34.
DCP					
21. Color platinum cobalt units or dominant wavelength, hue, luminance purity:					
Colorimetric (ADMI), or	110.1	2120 E			Note 18.
(Platinum cobalt), or	110.2	2120 B		I-1250-85	
Spectrophotometric	110.3	2120 C			
22. Copper—Total,[4] mg/L; Digestion [4] followed by:					
AA direct aspiration	220.1	3111 B or C	D1688-90(A or B)	I-3270-85 or I3271-85	974.27[3] p. 37.[9]
AA furnace	220.2	3113 B	D1688-90(C)		
ICP/AES [36]	[5]200.7	3120 B	D4190-82(88)		Note 34.
DCP [36] or					
Colorimetric (Neocuproine) or		3500-Cu D			Note 19.
(Bicinchoninate)		Or E			
23. Cyanide—Total, mg/L:					
Manual distillation with MgCl$_2$ followed by		4500-CN C	D2036-91(A)		
Titrimetric, or		4500-CN D			
Spectrophotometric, manual or		4500-CN E	D2036-91(A)	I-3300-85	p. 22.[9]
Automated [20]	[31]335.2				
24. Available Cyanide, mg/L:	[31]335.3				
Cyanide amenable to chlorination (CATC), Manual distillation with MgCl$_2$ followed by titrimetry or spectrophotometry.	335.1	4500-CN G	D2036-91(B)		
Flow injection and ligand exchange, followed by amperometry.					[44]OIA-1677
25. Fluoride—Total, mg/L:					
Manual distillation [6] followed by					
Electrode, manual or	340.2	4500-F B	D1179-93(B)		
Automated		4500-F C			
Colorimetric (SPADNS)	340.1	4500-F D	D1179-93(A)	I-4327-85	
Or Automated complexone	340.3	4500-F E			
26. Gold—Total,[4] mg/L; Digestion [4] followed by:					
AA direct aspiration	231.1	3111 B			
AA furnace, or	231.2				
DCP					Note 34.
27. Hardness—Total, as CaCO$_3$, mg/L					
Automated colorimetric,	130.1				

286

Environmental Protection Agency §136.3

Parameter	EPA	Standard Methods	ASTM	USGS/Other
Titrimetric (EDTA), or Ca plus Mg as their carbonates, by inductively coupled plasma or AA direct aspiration. (See Parameters 13 and 33).	130.2	2340 B or C	D1126-86(92)	
28. Hydrogen ion (pH), pH units				
Electrometric measurement, or	150.1	4500-H⁺ B	D1293-84(90)(A or B)	I-1586-85 .... 973.41.³
Automated electrode				Note 21.
29. Indium—Total,⁴ mg/L; Digestion⁴ followed by:				
AA direct aspiration or	235.1	3111 B		
AA furnace	235.2			
30. Iron—Total,⁴ mg/L; Digestion⁴ followed by:				
AA direct aspiration ³⁶	236.1	3111 B or C	D1068-90(A or B)	I-3381-85 .... 974.27.³
AA furnace	236.2	3113 B	D1068-90(C)	
ICP/AES³⁶	⁵200.7	3120 B		
DCP³⁶ or				Note 34.
Colorimetric (Phenanthroline)		3500—Fe D	D4190-82(88)	Note 22.
			D1068-90(D)	
31. Kjeldahl Nitrogen—Total, (as N), mg/L:				
Digestion and distillation followed by:				
Titration	351.3	4500—NH₃B or C	D3590-89(A)	973.48.³
Nesslerization	351.3	4500—NH₃E	D3590-89(A)	
Electrode	351.3	4500—NH₃C	D3590-89(A)	
Automated phenate colorimetric	351.3	4500—NH₃F or G		
Semi-automated block digester colorimetric	351.1			I-4551-78ₖ
Manual or block digester potentiometric	351.2		D3590-89(B)	
Block Digester, followed by:.	351.4		D3590-89(A)	
Auto distillation and Titration, or				Note 39.
Nesslerization				Note 40.
Flow injection gas diffusion				Note 41.
32. Lead—Total,⁴ mg/L; Digestion⁴ followed by:				
AA direct aspiration ³⁶	239.1	3111 B or C	D3559-90(A or B)	I-3399-85 .... 974.27.³
AA furnace	239.2	3113 B	D3559-90(D)	
ICP/AES ³⁶	⁵200.7	3120 B		
DCP³⁶			D4190-82(88)	Note 34.
Voltametry¹¹ or			D3559-90(C)	
Colorimetric (Dithizone)		3500—Pb D		
33. Magnesium—Total,⁴ mg/L; Digestion⁴ followed by:				
AA direct aspiration	242.1	3111 B	D511-93(B)	I-3447-85 .... 974.27.³
ICP/AES	⁵200.7	3120 B		
DCP, or				Note 34.
Gravimetric		3500—Mg D		
34. Manganese—Total,⁴ mg/L; Digestion⁴ followed by:				
AA direct aspiration ³⁶	243.1	3111 B	D858-90(A or B)	I-3454-85 .... 974.27.³
AA furnace	243.2	3113 B	D858-90(C)	
ICP/AES ³⁶	⁵200.7	3120 B		
DCP³⁶ or				Note 34.
Colorimetric (Persulfate), or		3500—Mn D	D4190-82(88)	920.203.³
(Periodate)				Note 23.
35. Mercury—Total,⁴ mg/L:				
Cold vapor, manual, or	245.1	3112 B	D3223-91	I-3462-85 .... ³977.22
Automated	245.2			

287

§ 136.3   40 CFR Ch. I (7-1-01 Edition)

TABLE IB.—LIST OF APPROVED INORGANIC TEST PROCEDURES—Continued

Parameter, units and method	EPA [1,35]	STD methods 18th ed.	ASTM	USGS [2]	Other
36. Molybdenum—Total,[4] mg/L; Digestion[4] followed by:	[43]1631				
AA direct aspiration	246.1	3111 D		I-3490-85	
AA furnace	246.2	3113 B			
ICP/AES	[5]200.7	3120 B			
DCP					Note 34.
37. Nickel—Total,[4] mg/L; Digestion[4] followed by:					
AA direct aspiration[36]	249.1	3111 B or C	D1886-90(A or B)	I-3499-85	
AA furnace	249.2	3113 B	D1886-90(C)		
ICP/AES[36]	[5]200.7	3120 B			
DCP[36], or			D4190-82(88)		Note 34.
Colorimetric (heptoxime)					
38. Nitrate (as N), mg/L:					
Colorimetric (Brucine sulfate), or Nitrate-nitrite N minus Nitrite N (See parameters 39 and 40).	352.1	3500-Ni D			973.50,[3] 419 D,[17] p. 28.[9]
39. Nitrate-nitrite (as N), mg/L:					
Cadmium reduction, Manual or	353.3	4500-NO₃ – E	D3867-90(B)	I-4545-85	
Automated, or	353.2	4500-NO₃ – F	D3867-90(A)		
Automated hydrazine	353.1	4500-NO₃ – H			
40. Nitrite (as N), mg/L: Spectrophotometric:					
Manual or	354.1	4500-NO₂ – B			Note 25.
Automated (Diazotization)					
41. Oil and grease—Total recoverable, mg/L:	413.1	5520 B[38]			
Gravimetric (extraction)					
Oil and grease and non-polar material, mg/L: Hexane extractable material (HEM): n-Hexane extraction and gravimetry[42]	1664, Rev. A				
Silica gel treated HEM (SGT-HEM): Silica gel treatment and gravimetry[42]	1664, Rev. A				
42. Organic carbon—Total (TOC), mg/L: Combustion or oxidation	415.1	5310 B, C, or D	D2579-93 (A or B)		973.47,[3] p. 14.[24]
43. Organic nitrogen (as N), mg/L: Total Kjeldahl N (Parameter 31) minus ammonia N (Parameter 4)					
44. Orthophosphate (as P), mg/L; Ascorbic acid method:					
Automated, or	365.1	4500-P F	D515-88(A)	I-4601-85	973.56.[3]
Manual single reagent	365.2	4500-P E			973.55.[3]
Manual two reagent	365.3				
45. Osmium—Total,[4] mg/L; Digestion[4] followed by:					
AA direct aspiration, or	252.1	3111 D			
AA furnace	252.2				
46. Oxygen, dissolved, mg/L:					
Winkler (Azide modification), or	360.2	4500-O C	D888-92(A)	I-1575-78[8]	973.45B.[3]

288

Environmental Protection Agency § 136.3

47. Electrode	360.1	4500–O G	D888–92(B)	
Palladium—Total,⁴ mg/L; Digestion⁴ followed by:				
AA direct aspiration, or	253.1	3111 B		p. S27.¹⁰
AA furnace	253.2			p. S28.¹⁰
DCP				Note 34.
48. Phenols, mg/L:				
Manual distillation²⁶	420.1			Note 27.
Followed by:				
Colorimetric (4AAP) manual, or	420.1			
Automated¹⁹	420.2			Note 27.
49. Phosphorus (elemental), mg/L:				
Gas-liquid chromatography				Note 28.
50. Phosphorus—Total, mg/L:				
Persulfate digestion followed by	365.2	4500–P B,5		973.55.³
Manual or	365.2 or 365.3	4500–P E		
Automated ascorbic acid reduction	365.1			
Semi-automated block digestor	365.4	4500–P F	I-4600–85	973.56.³
51. Platinum—Total,⁴ mg/L; Digestion⁴ followed by:				
AA direct aspiration	255.1	3111 B		
DCP	255.2			Note 34.
52. Potassium—Total,⁴ mg/L; Digestion⁴ followed by:				
AA direct aspiration	258.1	3111 B	I-3630–85	973.53.³
ICP/AES	⁵200.7	3120 B		
Flame photometric, or		3500–K D		
Colorimetric				317 B.¹⁷
53. Residue—Total, mg/L;				
Gravimetric, 103–105°	160.3	2540 B		I-3750–85
54. Residue—filterable, mg/L;				
Gravimetric, 180°	160.1	2540 C		I-1750–85
55. Residue—nonfilterable (TSS), mg/L;				
Gravimetric, 103–105° post washing of residue	160.2	2540 D		I-3765–85
56. Residue—settleable, mg/L;				
Volumetric, (Imhoff cone), or gravimetric	160.5	2540 F		
57. Residue—Volatile, mg/L;				
Gravimetric, 550°	160.4			I-3753–85
58. Rhodium—Total,⁴ mg/L; Digestion⁴ followed by:				
AA direct aspiration, or	265.1	3111 B		
AA furnace	265.2			
59. Ruthenium—Total,⁴ mg/L; Digestion⁴ followed by:				
AA direct aspiration, or	267.1	3111 B		
AA furnace	267.2			
60. Selenium—Total,⁴ mg/L; Digestion⁴ followed by:				
AA furnace	270.2	3113 B	D3859–93(B)	
ICP/AES,³⁶ or	⁵200.7	3120 B		
AA gaseous hydride		3114 B	D3859–93(A)	I-3667–85
61. Silica³⁷—Dissolved, mg/L; 0.45 micron filtration followed by:				
Colorimetric, Manual or	370.1	4500–Si D	D859–88	I-1700–85

§ 136.3   40 CFR Ch. I (7-1-01 Edition)

Table IB.—List of Approved Inorganic Test Procedures—Continued

Parameter, units and method	EPA[1,35]	STD methods 18th ed.	ASTM	USGS[2]	Other
Automated (Molybdosilicate), or	[5]200.7	3120 B		I-2700-85	
ICP					
62. Silver—Total,[4] mg/L; Digestion[4,29] followed by:					974.27,[3] p. 37.[9]
AA direct aspiration	272.1	3111 B or C		I-3720-85	
AA furnace	272.2	3113 B			
ICP/AES	[5]200.7	3120 B			
DCP					Note 34.
63. Sodium—Total,[4] mg/L; Digestion[4] followed by:					
AA direct aspiration	273.1	3111 B		I-3735-85	973.54.[3]
ICP/AES	[5]200.7	3120 B			
DCP, or					Note 34.
Flame photometric		3500 Na D			
64. Specific conductance, micromhos/cm at 25 °C:					
Wheatstone bridge	120.1	2510 B	D1125-91(A)	I-1780-85	973.40.[3]
65. Sulfate (as SO₄), mg/L:					
Automated colorimetric (barium chloranilate)	375.1				
Gravimetric, or	375.3	4500-SO₄⁻2 C or D			925.54,[3]
Turbidimetric	375.4		D516-90		426C.[30]
66. Sulfide (as S), mg/L:					
Titrimetric (iodine), or	376.1	4500-S ²E		I-3840-85	
Colorimetric (methylene blue)	376.2	4500-S ²D			
67. Sulfite (as SO₃), mg/L:					
Titrimetric (iodine-iodate)	377.1	4500-SO₃⁻2 B			
68. Surfactants, mg/L:					
Colorimetric (methylene blue)	425.1	5540 C	D2330-88		
69. Temperature, °C:					
Thermometric	170.1	2550 B			Note 32.
70. Thallium—Total,[4] mg/L; Digestion[4] followed by:					
AA direct aspiration	279.1	3111 B			
AA furnace	279.2				
ICP/AES, or	[5]200.7	3120 B			
71. Tin—Total,[4] mg/L; Digestion[4] followed by:					
AA direct aspiration	282.1	3111 B		I-3850-78[8]	
AA furnace, or	282.2	3113 B			
ICP/AES	[5]200.7				
72. Titanium—Total,[4] mg/L; Digestion[4] followed by:					
AA direct aspiration	283.1	3111 D			
AA furnace	283.2				Note 34.
DCP					
73. Turbidity, NTU:					
Nephelometric	180.1	2130 B	D1889-88(A)	I-3860-85	
74. Vanadium—Total,[4] mg/L; Digestion[4] followed by:					
AA direct aspiration	286.1	3111 D	D3373-93		
AA furnace	286.2				

290

Environmental Protection Agency § 136.3

ICP/AES	[5]200.7	3120 B			Note 34.
DCP, or					
Colorimetric (Gallic acid)		3500–V D			
75. Zinc—Total,[4] mg/L; Digestion[4] followed by:					
AA direct aspiration[36]	289.1	3111 B or C	D4190-82(88)	I-3900-85	974.27,[3] p. 37.[9]
AA furnace	289.2				
ICP/AES[36]	[5]200.7	3120 B	D1691-90 (A or B)		
DCP,[36] or		3500–Zn E			Note 34.
Colorimetric (Dithizone) or					
(Zincon)		3500–Zn F	D4190-82(88)		Note 33.

Table IB Notes:

[1] "Methods for Chemical Analysis of Water and Wastes", Environmental Protection Agency, Environmental Monitoring Systems Laboratory–Cincinnati (EMSL-CI), EPA-600/4-79-020, Revised March 1983 and 1979 where applicable.

[2] Fishman, M.J., et al., "Methods for Analysis of Inorganic Substances in Water and Fluvial Sediments," U.S. Department of the Interior, Techniques of Water—Resource Investigations of the U.S. Geological Survey, Denver, CO, Revised 1989, unless otherwise stated.

[3] "Official Methods of Analysis of the Association of Official Analytical Chemists," methods manual, 15th ed. (1990).

[4] For the determination of total metals the sample is not filtered before processing. A digestion procedure is required to solubilize suspended material and to destroy possible organic-metal complexes. Two digestion procedures are given in "Methods for Chemical Analysis of Water and Wastes, 1979 and 1983". One (section 4.1.3), is a vigorous digestion using nitric acid. A less vigorous digestion using nitric and hydrochloric acids (section 4.1.4) is preferred; however, the analyst should be cautioned that this mild digestion may not suffice for all samples types. Particularly, if a colorimetric procedure is to be employed, it is necessary to ensure that all organo-metallic bonds be broken so that the metal is in a reactive state. In those situations, the vigorous digestion is to be preferred making certain that, at no time does the sample go to dryness. Samples containing large amounts of organic materials may also benefit by this vigorous digestion, however, vigorous digestion with concentrated nitric acid will convert antimony and tin to insoluble oxides and render them unavailable for analysis. Use of ICP/AES as well as determinations for certain elements such as antimony, arsenic, the noble metals, mercury, selenium, silver, tin, and titanium require a modified sample digestion procedure and in all cases the method write-up should be consulted for specific instructions and/or cautions.

NOTE TO TABLE IB NOTE 4: If the digestion procedure for direct aspiration AA included in one of the other approved references is different than the above, the EPA procedure must be used.

Dissolved metals are defined as those constituents which will pass through a 0.45 micron membrane filter. Following filtration of the sample, the referenced procedure for total metals must be followed. Sample digestion of the filtrate for dissolved metals (or digestion of the original sample solution for total metals) may be omitted for AA (direct aspiration or graphite furnace) and ICP analyses, provided the sample solution to be analyzed meets the following criteria:

a. has a low COD (<20)
b. is visibly transparent with a turbidity measurement of 1 NTU or less
c. is colorless with no perceptible odor, and
d. is of one liquid phase and free of particulate or suspended matter following acidification.

[5] The full text of Method 200.7, "Inductively Coupled Plasma Atomic Emission Spectrometric Method for Trace Element Analysis of Water and Wastes," is given at Appendix C of this Part 136.

[6] Manual distillation is not required if comparability data on representative effluent samples are on company file to show that this preliminary distillation step is not necessary; however, manual distillation will be required to resolve any controversies.

[7] Ammonia, Automated Electrode Method, Industrial Method Number 379–75 WE, dated February 19, 1976, (Bran & Luebbe (Technicon) Auto Analyzer II, Bran & Luebbe Analyzing Technologies, Inc., Elmsford, NY 10523.

[8] The approved method is that cited in "Methods for Determination of Inorganic Substances in Water and Fluvial Sediments", USGS TWRI, Book 5, Chapter A1 (1979).

[9] American National Standard on Photographic Processing Effluents, Apr. 2, 1975. Available from ANSI, 1430 Broadway, New York, NY 10018.

[10] "Selected Analytical Methods Approved and Cited by the United States Environmental Protection Agency", Supplement to the Fifteenth Edition of Standard Methods for the Examination of Water and Wastewater (1981).

[11] The use of normal and differential pulse voltage ramps to increase sensitivity and resolution is acceptable.

[12] Carbonaceous biochemical oxygen demand ($CBOD_5$) must not be confused with the traditional BOD_5 test which measures "total BOD." The addition of the nitrification inhibitor is not a procedural option, but must be included to report the $CBOD_5$ parameter. A discharger whose permit requires reporting the traditional BOD_5, may not use a nitrification inhibitor in the procedure for reporting the results. Only when a discharger's permit specifically states $CBOD_5$, is required can the permittee report data using the nitrification inhibitor.

[13] OIC Chemical Oxygen Demand Method, Oceanography International Corporation, 1978, 512 West Loop, P.O. Box 2980, College Station, TX 77840.

[14] Chemical Oxygen Demand, Method 8000, Hach Handbook of Water Analysis, 1979, Hach Chemical Company, P.O. Box 389, Loveland, CO 80537.

[15] The back titration method will be used to resolve controversy.

[16] Orion Research Instruction Manual, Residual Chlorine Electrode Model 97–70, 1977, Orion Research Incorporated, 840 Memorial Drive, Cambridge, MA 02138. The calibration graph for the Orion residual chlorine method must be derived using a reagent blank and three standard solutions, containing 0.2, 1.0, and 5.0 ml 0.00281 N potassium iodate/100 ml solution, respectively.

[17] The approved method is that cited in Standard Methods for the Examination of Water and Wastewater, 14th Edition, 1976.

[18] National Council of the Paper Industry for Air and Stream Improvement, (Inc.) Technical Bulletin 253, December 1971.

[19] Copper, Biocinchoinate Method, Method 8506, Hach Handbook of Water Analysis, 1979, Hach Chemical Company, P.O. Box 389, Loveland, CO 80537.

§ 136.3 40 CFR Ch. I (7–1–01 Edition)

[20] After the manual distillation is completed, the autoanalyzer manifolds in EPA Methods 335.3 (cyanide) or 420.2 (phenols) are simplified by connecting the re-sample line directly to the sampler. When using the manifold setup shown in Method 335.3, the buffer 6.2 should be replaced with the buffer 7.6 found in Method 335.2.
[21] Hydrogen ion (pH) Automated Electrode Method, Industrial Method Number 378–75WA, October 1976, Bran & Luebbe (Technicon) Autoanalyzer II. Bran & Luebbe Analyzing Technologies, Inc., Elmsford, NY 10523.
[22] Iron, 1,10-Phenanthroline Method, Method 8008, 1980, Hach Chemical Company, P.O. Box 389, Loveland, CO 80537.
[23] Manganese, Periodate Oxidation Method, Method 8034, Hach Handbook of Wastewater Analysis, 1979, pages 2–113 and 2–117, Hach Chemical Company, Loveland, CO 80537.
[24] Wershaw, R.L., et al. "Methods for Analysis of Organic Substances in Water," Techniques of Water-Resources Investigation of the U.S. Geological Survey, Book 5, Chapter A3, (1972 Revised 1987) p. 14.
[25] Nitrogen, Nitrite, Method 8507, Hach Chemical Company, P.O. Box 389, Loveland, CO 80537.
[26] Just prior to distillation, adjust the sulfuric-acid-preserved sample to pH 4 with 1 + 9 NaOH.
[27] The approved method is cited in Standard Methods for the Examination of Water and Wastewater, 14th Edition. The colorimetric reaction is conducted at a pH of 10.0±0.2. The approved methods are given on pp 576–81 of the 14th Edition; Method 510A for distillation, Method 510B for the manual colorimetric procedure, or Method 510C for the manual spectophotometric procedure.
[28] R. F. Addison and R.G. Ackman, "Direct Determination of Elemental Phosphorus by Gas-Liquid Chromatography," Journal of Chromatography, vol. 47, No. 3, pp. 421–426, 1970.
[29] Approved methods for the analysis of silver in industrial wastewaters at concentrations of 1 mg/L are inadequate where silver exists as an inorganic halide. Silver halides such as the bromide and chloride are relatively insoluble in reagents such as nitric acid but are readily soluble in an aqueous buffer of sodium thiosulfate and sodium hydroxide to pH of 12. Therefore, for levels of silver above 1 mg/L, 20 mL of sample should be diluted 40 mL each of 2 M Na₂S₂O₃ and NaOH. Standards should be prepared in the same manner. For levels of silver below 1 mg/L the approved method is satisfactory.
[30] The approved method is that cited in Standard Methods for the Examination of Water and Wastewater, 15th Edition.
[31] EPA Methods 335.2 and 335.3 require the NaOH absorber solution final concentration to be adjusted to 0.25 N before colorimetric determination of total cyanide.
[32] Stevens, H.H., Ficke, J.F., and Smoot, G.F., "Water Temperature—Influential Factors, Field Measurement and Data Presentation", Techniques of Water-Resources Investigations of the U.S. Geological Survey, Book 1, Chapter D1, 1975.
[33] Zinc, Zincon Method, Method 8009, Hach Handbook of Water Analysis, 1979, pages 2–231 and 2–333, Hach Chemical Company, Loveland, CO 80537.
[34] "Direct Current Plasma (DCP) Optical Emission Spectrometric Method for Trace Elemental Analysis of Water and Wastes, Method AES0029," 1986—Revised 1991, Fison Instruments, Inc., 32 Commerce Center, Cherry Hill Drive, Danvers, MA 01923.
[35] Precision and recovery statements for the atomic absorption direct aspiration and graphite furnace methods, and for the spectrophotometric SDDC method for arsenic are provided in Appendix D of this part titled, "Precision and Recovery Statements for Methods for Measuring Metals".
[36] "Closed Vessel Microwave Digestion of Wastewater Samples for Determination of Metals", CEM Corporation, P.O. Box 200, Matthews, NC 28106–0200, April 16, 1992. Available from the CEM Corporation.
[37] When determining boron and silica, only plastic, PTFE, or quartz laboratory ware may be used from start until completion of analysis.
[38] Only the trichlorofluoromethane extraction solvent is approved.
[39] Nitrogen, Total Kjeldahl, Method PAI–DK01 (Block Digestion, Steam Distillation, Titrimetric Detection), revised 12/22/94, Perstop Analytical Corporation.
[40] Nitrogen, Total Kjeldahl, Method PAI–DK02 (Block Digestion, Steam Distillation, Colorimetric Detection), revised 12/22/94, Perstop Analytical Corporation.
[41] Nitrogen, Total Kjeldahl, Method PAI–DK03 (Block Digestion, Automated FIA Gas Diffusion), revised 12/22/94, Perstop Analytical Corporation.
[42] Method 1664, Revision A "n-Hexane Extractable Material (HEM; Oil and Grease) and Silica Gel Treated n-Hexane Extractabike Material (SGT–HEM; Non-polar Material) by Extraction and Gravimetry" EPA-821–R–98–002, February 1999. Available at NTIS, PB–121949, U.S. Department of Commerce, 5285 Port Royal, Springfield, Virginia 22161.
[43] The application of clean techniques described in EPA's draft Method 1669: Sampling Ambient Water for Trace Metals at EPA Water Quality Criteria Levels (EPA-821–R–96–011) are recommended to preclude contamination at low-level, trace metal determinations.
[44] Available Cyanide, Method OIA–1677 (Available Cyanide by Flow Injection, Ligand Exchange, and Amperometry), ALPKEM, A Division of OI Analytical, P.O. Box 9010, College Station, TX 77842–9010.

TABLE IC.—LIST OF APPROVED TEST PROCEDURES FOR NON-PESTICIDE ORGANIC COMPOUNDS

Parameter[1]	EPA method number[2 7]					
	GC	GC/MS	HPLC	Standard method 18th Ed.	ASTM	Other
1. Acenaphthene	610	625, 1625	610	6410 B, 6440 B	D4657-92	
2. Acenaphthylene	610	625, 1625	610	6410 B, 6440 B	D4657-92	
3. Acrolein	603	[4]604, 1624				
4. Acrylonitrile	603	[4]624, 1624	610			
5. Anthracene	610	625, 1625	610	6410 B, 6440 B	D4657-92	
6. Benzene	602	624, 1624		6210 B, 6220 B		
7. Benzidine		[5]625, 1625	605			Note 3, p.1.
8. Benzo(a)anthracene	610	625, 1625	610	6410 B, 6440 B	D4657-92	
9. Benzo(a)pyrene	610	625, 1625	610	6410 B, 6440 B	D4657-92	

Environmental Protection Agency § 136.3

10. Benzo(b)fluoranthene	610	625, 1625	610	6410 B, 6440 B	D4657-92
11. Benzo(g, h, i)perylene	610	625, 1625	610	6410 B, 6440 B	D4657-92
12. Benzo(k)fluoranthene	610	625, 1625	610	6410 B, 6440 B	D4657-92
13. Benzyl chloride					Note 3, p.130; Note 6, p. S102.
14. Benzyl butyl phthalate	606	625, 1625		6410 B	
15. Bis(2-chloroethoxy) methane	611	625, 1625		6410 B	
16. Bis(2-chloroethyl) ether	611	625, 1625		6410 B, 6230 B	
17. Bis (2-ethylhexyl) phthalate	606	625, 1625		6410 B	
18. Bromodichloromethane	601	624, 1624		6210 B, 6230 B	
19. Bromoform	601	624, 1624		6210 B, 6230 B	
20. Bromomethane	601	624, 1624		6210 B, 6230 B	
21. 4-Bromophenylphenyl ether	611	625, 1625		6410 B	
22. Carbon tetrachloride	601	624, 1624		6230 B, 6410 B	Note 3, p.130.
23. 4-Chloro-3-methylphenol	604	625, 1625		6410 B, 6420 B	Note 3, p.130.
24. Chlorobenzene	601, 602	624, 1624		6210 B, 6220 B	
25. Chloroethane	601	624, 1624		6230 B	
26. 2-Chloroethylvinyl ether	601	624, 1624		6210 B, 6230 B	Note, p.130.
27. Chloroform	601	624, 1624		6210 B, 6230 B	
28. Chloromethane	601	624, 1624		6210 B, 6230 B	
29. 2-Chloronaphthalene	612	625, 1625		6410 B	
30. 2-Chlorophenol	604	625, 1625		6410 B, 6420 B	
31. 4-Chlorophenylphenyl ether	611	625, 1625		6410 B	
32. Chrysene	610	625, 1625	610	6410 B, 6440 B	D4657-92
33. Dibenzo(a,h)anthracene	610	625, 1625	610	6410 B, 6440 B	D4657-92
34. Dibromochloromethane	601	624, 1624		6210 B, 6230 B	
35. 1, 2-Dichlorobenzene	601,602,612	624,625,1625		6410 B, 6230 B, 6220 B	
36. 1, 3-Dichlorobenzene	601,602,612	624,625,1625		6410 B, 6230 B, 6220 B	
37. 1,4-Dichlorobenzene	601, 602, 612	624, 625, 1625		6410 B, 6220 B, 6230 B	
38. 3, 3-Dichlorobenzidine		625, 1625	605	6410 B	
39. Dichlorodifluoromethane	601			6230 B	
40. 1, 1-Dichloroethane	601	624, 1624		6230 B, 6210 B	
41. 1, 2-Dichloroethane	601	624, 1624		6210 B, 6210 B	
42. 1, 1-Dichloroethene	601	624, 1624		6230 B, 6210 B	
43. trans-1, 2-Dichloroethene	601	624, 1624		6210 B, 6210 B	
44. 2, 4-Dichlorophenol	604	625, 1625		6420 B, 6410 B	
45. 1, 2-Dichloropropane	601	624, 1624		6230 B, 6210 B	
46. cis-1, 3-Dichloropropene	601	624, 1624		6230 B, 6210 B	
47. trans-1, 3-Dichloropropene	601	624, 1624		6230 B, 6210 B	
48. Diethyl phthalate	606	625, 1625		6410 B	
49. 2, 4-Dimethylphenol	604	625, 1625		6420 B, 6410 B	
50. Dimethyl phthalate	606	625, 1625		6410 B	
51. Di-n-butyl phthalate	606	625, 1625		6410 B	
52. Di-n-octyl phthalate	606	625, 1625		6410 B	
53. 2,4-Dinitrophenol	604	625, 1625		6420 B, 6410 B	
54. 2,4-Dinitrotoluene	609	625, 1625		6410 B	
55. 2, 6-Dinitrotoluene	609	625, 1625		6410 B	

TABLE IC.—LIST OF APPROVED TEST PROCEDURES FOR NON-PESTICIDE ORGANIC COMPOUNDS—Continued

Parameter [1]	GC	GC/MS	HPLC	EPA method number [2,7] Standard method 18th Ed.	ASTM	Other
56. Epichlorohydrin						Note 3, p. 130
57. Ethylbenzene	602	624, 1624		6220 B, 6210 B		
58. Fluoranthene	610	625, 1625	610	6410 B, 6440 B	D4657-92	Note 6, p.S102.
59. Fluorene	610	625, 1625	610	6410 B, 6440 B	D4657-92	
60. 1,2,3,4,6,7,8-Heptachlorodibenzofuran		1613				
61. 1,2,3,4,7,8,9-Heptachlorodibenzofuran		1613				
62. 1,2,3,4,6,7,8-Heptachlorodibenzo-p-dioxin		1613				
63. Hexachlorobenzene	612	625, 1625		6410 B		
64. Hexachlorobutadiene	612	625, 1625		6410 B		
65. Hexachlorocyclopentadiene	612	625, 1625 [5]		6410 B		
66. 1,2,3,4,7,8-Hexachlorodibenzofuran		1613				
67. 1,2,3,6,7,8-Hexachlorodibenzofuran		1613				
68. 1,2,3,7,8,9-Hexachlorodibenzofuran		1613				
69. 2,3,4,6,7,8-Hexachlorodibenzofuran		1613				
70. 1,2,3,4,7,8-Hexachlorodibenzo-p-dioxin		1613				
71. 1,2,3,6,7,8-Hexachlorodibenzo-p-dioxin		1613				
72. 1,2,3,7,8,9-Hexachlorodibenzo-p-dioxin		1613				
73. Hexachloroethane	612	625, 1625		6410 B		
74. Indeno(1,2,3-cd)pyrene	610	625, 1625	610	6410 B, 6440 B	D4657-87	
75. Isophorone	609	625, 1625		6410 B		Note 3, p. 130.
76. Methylene chloride	601	624, 1624		6230 B		
77. 2-Methyl-4,6-dinitrophenol	604	625, 1625		6420 B, 6410 B		
78. Naphthalene	610	625, 1625	610	6410 B, 6440 B	D4657-87	
79. Nitrobenzene	609	625, 1625		6410 B		
80. 2-Nitrophenol	604	625, 1625		6410 B, 6420 B		
81. 4-Nitrophenol	604	625, 1625		6410 B, 6420 B		
82. N-Nitrosodimethylamine	607	625, 1625		6410 B		
83. N-Nitrosodi-n-propylamine	607	625, 1625 [5]		6410 B		
84. N-Nitrosodiphenylamine	607	625, 1625 [5]		6410 B		
85. Octachlorodibenzofuran		1613				
86. Octachlorodibenzo-p-dioxin		1613				
87. 2,2-Oxybis(1-chloropropane)	611	625, 1625		6410 B		
88. PCB-1016	608	625		6410 B		Note 3, p. 43.
89. PCB-1221	608	625		6410 B		Note 3, p. 43.
90. PCB-1232	608	625		6410 B		Note 3, p. 43.
91. PCB-1242	608	625		6410 B		Note 3, p. 43.
92. PCB-1248	608	625		6410 B		
93. PCB-1254	608	625		6410 B		Note 3, p. 43.
94. PCB-1260	608	625		6410 B, 6630 B		Note 3, p. 43.
95. 1,2,3,7,8-Pentachlorodibenzofuran		1613				
96. 2,3,4,7,8-Pentachlorodibenzofuran		1613				
97. 1,2,3,7,8-Pentachlorodibenzo-p-dioxin		1613				

Environmental Protection Agency §136.3

98. Pentachlorophenol	604	625, 1625	6410 B, 6630 B		Note 3, p. 140.
99. Phenanthrene	610	625, 1625	6410 B, 6440 B	610	
100. Phenol	604	625, 1625	6420 B, 6410 B		
101. Pyrene	610	625, 1625	6410 B, 6440 B	610	
102. 2,3,7,8-Tetrachlorodibenzofuran		1613			
103. 2,3,7,8-Tetrachlorodibenzo-p-dioxin		613, 1613[5]			
104. 1,1,2,2-Tetrachloroethane	601	624, 1624	6230 B, 6210 B		Note 3, p. 130.
105. Tetrachloroethene	601	624, 1624	6230 B, 6410 B		Note 3, p. 130.
106. Toluene	602	624, 1624	6210 B, 6220 B		
107. 1,2,4-Trichlorobenzene	612	625, 1625	6410 B		Note 3, p. 130.
108. 1,1,1-Trichloroethane	601	624, 1624	6210 B, 6230 B		
109. 1,1,2-Trichloroethane	601	624, 1624	6210 B, 6230 B		Note 3, p. 130.
110. Trichloroethene	601	624, 1624	6210 B, 6230 B		
111. Trichlorofluoromethane	601	624	6410 B, 6240 B		
112. 2,4,6-Trichlorophenol	604	625, 1625	6210 B, 6230 B		
113. Vinyl chloride	601	624, 1624			

Table 1C notes:

[1] All parameters are expressed in micrograms per liter (μg/L) except for Method 1613 in which the parameters are expressed in picograms per liter (pg/L).

[2] The full text of Methods 601–613, 624, 625, 1624, and 1625, are given at appendix A, "Test Procedures for Analysis of Organic Pollutants," of this part 136. The full text of Method 1613 is incorporated by reference into this part 136 and is available from the National Technical Information Services as stock number PB95–104774. The standardized test procedure to be used to determine the method detection limit (MDL) for these test procedures is given at appendix B, "Definition and Procedures for the Determination of the Method Detection Limit," of this part 136.

[3] "Methods for Benzidine: Chlorinated Organic Compounds, Pentachlorophenol and Pesticides in Water and Wastewater," U.S. Environmental Protection Agency, September, 1978.

[4] Method 624 may be extended to screen samples for Acrolein and Acrylonitrile. However, when they are known to be present, the preferred method for these two compounds is Method 603 or Method 1624.

[5] Method 625 may be extended to include benzidine, hexachlorocyclopentadiene, N-nitrosodimethylamine, and N-nitrosodiphenylamine. However, when they are known to be present, Methods 605, 607, and 612, or Method 1625, are preferred methods for these compounds.

[6] 625, Screening only.

[7] "Selected Analytical Methods Approved and Cited by the United States Environmental Protection Agency", Supplement to the Fifteenth Edition of Standard Methods for the Examination of Water and Wastewater (1981).

[7] Each Analyst must make an initial, one-time demonstration of their ability to generate acceptable precision and accuracy with Methods 601–603, 624, 625, 1624, and 1625 (See Appendix A of this Part 136) in accordance with procedures each in section 8.2 of each of these Methods. Additionally, each laboratory, on an on-going basis must spike and analyze 10% (5% for Methods 624 and 625 and 100% for methods 1624 and 1625) of all samples to monitor and evaluate laboratory data quality in accordance with sections 8.3 and 8.4 of these Methods. When the recovery of any parameter falls outside the warning limits, the analytical results for that parameter in the unspiked sample are suspect and cannot be reported to demonstrate regulatory compliance.

NOTE: These warning limits are promulgated as an "interim final action with a request for comments."

[8] "Organochlorine Pesticides and PCBs in Wastewater Using Empore TM Disk", 3M Corporation Revised 10/28/94.

TABLE ID.—LIST OF APPROVED TEST PROCEDURES FOR PESTICIDES [1]

Parameter	Method	EPA[27]	Standard methods 18th Ed.	ASTM	Other
1. Aldrin	GC	608	6630 B & C	D3086-90	Note 3, p. 7; note 4, p. 30; note 8.
	GC/MS	625	6410 B		Note 3, p. 83; Note 6, p. S68.
2. Ametryn	GC				Note 3, p. 94; Note 6, p. S16.
3. Aminocarb	TLC				Note 3, p. 83; Note 6, p. S68.
4. Atraton	GC				Note 3, p. 83; Note 6, p. S68.
5. Atrazine	GC				Note 3, p. 25; Note 6, p. S51.
6. Azinphos methyl	GC				Note 3, p. 104; Note 6, p. S64.
7. Barban	TLC				Note 3, p. 7; note 8.
8. α-BHC	GC	608	6630 B & C	D3086-90	
	GC/MS	[5]625	6410 B		

295

Environmental Protection Agency § 136.3

(43) Method OIA–1677, Available Cyanide by Flow Injection, Ligand Exchange, and Amperometry. August 1999. ALPKEM, OI Analytical, Box 648, Wilsonville, Oregon 97070 (EPA–821–R–99–013). Available from: National Technical Information Service, 5285 Port Royal Road, Springfield, Virginia 22161. Publication No. PB99–132011. Cost: $22.50. Table IB, Note 44.

(c) Under certain circumstances the Regional Administrator or the Director in the Region or State where the discharge will occur may determine for a particular discharge that additional parameters or pollutants must be reported. Under such circumstances, additional test procedures for analysis of pollutants may be specified by the Regional Administrator, or the Director upon the recommendation of the Director of the Environmental Monitoring Systems Laboratory—Cincinnati.

(d) Under certain circumstances, the Administrator may approve, upon recommendation by the Director, Environmental Monitoring Systems Laboratory—Cincinnati, additional alternate test procedures for nationwide use.

(e) Sample preservation procedures, container materials, and maximum allowable holding times for parameters cited in Tables IA, IB, IC, ID, and IE are prescribed in Table II. Any person may apply for a variance from the prescribed preservation techniques, container materials, and maximum holding times applicable to samples taken from a specific discharge. Applications for variances may be made by letters to the Regional Administrator in the Region in which the discharge will occur. Sufficient data should be provided to assure such variance does not adversely affect the integrity of the sample. Such data will be forwarded, by the Regional Administrator, to the Director of the Environmental Monitoring Systems Laboratory—Cincinnati, Ohio for technical review and recommendations for action on the variance application. Upon receipt of the recommendations from the Director of the Environmental Monitoring Systems Laboratory, the Regional Administrator may grant a variance applicable to the specific charge to the applicant. A decision to approve or deny a variance will be made within 90 days of receipt of the application by the Regional Administrator.

TABLE II—REQUIRED CONTAINERS, PRESERVATION TECHNIQUES, AND HOLDING TIMES

Parameter No./name	Container [1]	Preservation [2,3]	Maximum holding time [4]
Table IA—Bacteria Tests:			
1–4 Coliform, fecal and total	P,G	Cool, 4C, 0.008% $Na_2S_2O_3$ [5]	6 hours.
5 Fecal streptococci	P,G	Cool, 4C, 0.008% $Na_2S_2O_3$ [5]	6 hours.
Table IA—Aquatic Toxicity Tests:			
6–10 Toxicity, acute and chronic	P,G	Cool, 4 °C [16]	36 hours.
Table IB—Inorganic Tests:			
1. Acidity	P, G	Cool, 4°C	14 days.
2. Alkalinity	P, G	...do	Do.
4. Ammonia	P, G	Cool, 4°C, H_2SO_4 to pH<2	28 days.
9. Biochemical oxygen demand	P, G	Cool, 4°C	48 hours.
10. Boron	P, PFTE, or Quartz.	HNO_3 TO pH<2	6 months.
11. Bromide	P, G	None required	28 days.
14. Biochemical oxygen demand, carbonaceous	P, G	Cool, 4°C	48 hours.
15. Chemical oxygen demand	P, G	Cool, 4°C, H_2SO_4 to pH<2	28 days.
16. Chloride	P, G	None required	Do.
17. Chlorine, total residual	P, G	...do	Analyze immediately.
21. Color	P, G	Cool, 4°C	48 hours.
23–24. Cyanide, total and amenable to chlorination.	P, G	Cool, 4°C, NaOH to pH>12, 0.6g ascorbic acid [5].	14 days.[6]
25. Fluoride	P	None required	28 days.
27. Hardness	P, G	HNO_3 to pH<2, H_2SO_4 to pH<2	6 months.
28. Hydrogen ion (pH)	P, G	None required	Analyze immediately.
31, 43. Kjeldahl and organic nitrogen	P, G	Cool, 4°C, H_2SO_4 to pH<2	28 days.
Metals:[7]			
18. Chromium VI	P, G	Cool, 4°C	24 hours.
35. Mercury	P, G	HNO_3 to pH<2	28 days.

§ 136.3

40 CFR Ch. I (7-1-01 Edition)

TABLE II—REQUIRED CONTAINERS, PRESERVATION TECHNIQUES, AND HOLDING TIMES—Continued

Parameter No./name	Container [1]	Preservation [2,3]	Maximum holding time [4]
3, 5–8, 12, 13, 19, 20, 22, 26, 29, 30, 32–34, 36, 37, 45, 47, 51, 52, 58–60, 62, 63, 70–72, 74, 75. Metals, except boron, chromium VI and mercury.	P, G	......do......	6 months.
38. Nitrate	P, G	Cool, 4°C	48 hours.
39. Nitrate-nitrite	P, G	Cool, 4°C, H_2SO_4 to pH<2	28 days.
40. Nitrite	P, G	Cool, 4°C	48 hours.
41. Oil and grease	G	Cool to 4°C, HCl or H_2SO_4 to pH<2.	28 days.
42. Organic Carbon	P, G	Cool to 4 °C HCl or H_2SO4 or H_3PO4, to pH<2.	28 days.
44. Orthophosphate	P, G	Filter immediately, Cool, 4°C	48 hours.
46. Oxygen, Dissolved Probe	G Bottle and top.	None required	Analyze immediately.
47. Winkler	......do	Fix on site and store in dark	8 hours.
48. Phenols	G only	Cool, 4°C, H_2SO_4 to pH<2	28 days.
49. Phosphorus (elemental)	G	Cool, 4°C	48 hours.
50. Phosphorus, total	P, G	Cool, 4°C, H_2SO_4 to pH<2	28 days.
53. Residue, total	P, G	Cool, 4°C	7 days.
54. Residue, Filterable	P, G	......do	7 days.
55. Residue, Nonfilterable (TSS)	P, G	......do	7 days.
56. Residue, Settleable	P, G	......do	48 hours.
57. Residue, volatile	P, G	......do	7 days.
61. Silica	P, PFTE, or Quartz.	Cool, 4 °C	28 days.
64. Specific conductance	P, G	......do	Do.
65. Sulfate	P, G	......do	Do.
66. Sulfide	P, G	Cool, 4°C add zinc acetate plus sodium hydroxide to pH>9.	7 days.
67. Sulfite	P, G	None required	Analyze immediately.
68. Surfactants	P, G	Cool, 4°C	48 hours.
69. Temperature	P, G	None required	Analyze.
73. Turbidity	P, G	Cool, 4°C	48 hours.
Table IC—Organic Tests [8]			
13, 18–20, 22, 24–28, 34–37, 39–43, 45–47, 56, 76, 104, 105, 108–111, 113. Purgeable Halocarbons.	G, Teflon-lined septum.	Cool, 4 °C, 0.008% $Na_2S_2O_3$ [5].	14 days.
6, 57, 106. Purgeable aromatic hydrocarbons	......do	Cool, 4 °C, 0.008% $Na_2S_2O_3$,[5] HCl to pH2[9].	Do.
3, 4. Acrolein and acrylonitrile	......do	Cool, 4 °C, 0.008% $Na_2S_2O_3$,[5] adjust pH to 4–5[10].	Do.
23, 30, 44, 49, 53, 77, 80, 81, 98, 100, 112. Phenols [11].	G, Teflon-lined cap..	Cool, 4 °C, 0.008% $Na_2S_2O_3$ [5]	7 days until extraction; 40 days after extraction.
7, 38. Benzidines [11]	......do	......do	7 days until extraction.[13]
14, 17, 48, 50–52. Phthalate esters [11]	......do	Cool, 4 °C	7 days until extraction; 40 days after extraction.
82–84. Nitrosamines [11][14]	......do	Cool, 4 °C, 0.008% $Na_2S_2O_3$,[5] store in dark.	Do.
88–94. PCBs [11]	......do	Cool, 4 °C	Do.
54, 55, 75, 79. Nitroaromatics and isophorone [11]	......do	Cool, 4 °C, 0.008% $Na_2S_2O_3$,[5] store in dark.	Do.
1, 2, 5, 8–12, 32, 33, 58, 59, 74, 78, 99, 101. Polynuclear aromatic hydrocarbons [11].	......do	......do	Do.
15, 16, 21, 31, 87. Haloethers [11]	......do	Cool, 4 °C, 0.008% $Na_2S_2O_3$ [5]	Do.
29, 35–37, 63–65, 73, 107. Chlorinated hydrocarbons [11].	......do	Cool, 4 °C	Do.
60–62, 66–72, 85, 86, 95–97, 102, 103. CDDs/CDFs [11].			
aqueous: field and lab preservation.	G	Cool, 0–4 °C, pH<9, 0.008% $Na_2S_2O_3$ [5].	1 year.
Solids, mixed phase, and tissue: field preservation..	......do	Cool, <4 °C	7 days.
Solids, mixed phase, and tissue: lab preservation	......do	Freeze, <−10 °C	1 year.
Table ID—Pesticides Tests:			
1–70. Pesticides [11]	......do	Cool, 4°C, pH 5–9 [15]	Do.
Table IE—Radiological Tests:			
1–5. Alpha, beta and radium	P, G	HNO_3 to pH<2	6 months.

Table II Notes

Environmental Protection Agency § 136.4

[1] Polyethylene (P) or glass (G). For microbiology, plastic sample containers must be made of sterilizable materials (polypropylene or other autoclavable plastic).

[2] Sample preservation should be performed immediately upon sample collection. For composite chemical samples each aliquot should be preserved at the time of collection. When use of an automated sampler makes it impossible to preserve each aliquot, then chemical samples may be preserved by maintaining at 4°C until compositing and sample splitting is completed.

[3] When any sample is to be shipped by common carrier or sent through the United States Mails, it must comply with the Department of Transportation Hazardous Materials Regulations (49 CFR part 172). The person offering such material for transportation is responsible for ensuring such compliance. For the preservation requirements of Table II, the Office of Hazardous Materials, Materials Transportation Bureau, Department of Transportation has determined that the Hazardous Materials Regulations do not apply to the following materials: Hydrochloric acid (HCl) in water solutions at concentrations of 0.04% by weight or less (pH about 1.96 or greater); Nitric acid (HNO_3) in water solutions at concentrations of 0.15% by weight or less (pH about 1.62 or greater); Sulfuric acid (H_2SO_4) in water solutions at concentrations of 0.35% by weight or less (pH about 1.15 or greater); and Sodium hydroxide (NaOH) in water solutions at concentrations of 0.080% by weight or less (pH about 12.30 or less).

[4] Samples should be analyzed as soon as possible after collection. The times listed are the maximum times that samples may be held before analysis and still be considered valid. Samples may be held for longer periods only if the permittee, or monitoring laboratory, has data on file to show that for the specific types of samples under study, the analytes are stable for the longer time, and has received a variance from the Regional Administrator under § 136.3(e). Some samples may not be stable for the maximum time period given in the table. A permittee, or monitoring laboratory, is obligated to hold the sample for a shorter time if knowledge exists to show that this is necessary to maintain sample stability. See § 136.3(e) for details. The term "analyze immediately" usually means within 15 minutes or less of sample collection.

[5] Should only be used in the presence of residual chlorine.

[6] Maximum holding time is 24 hours when sulfide is present. Optionally all samples may be tested with lead acetate paper before pH adjustments in order to determine if sulfide is present. If sulfide is present, it can be removed by the addition of cadmium nitrate powder until a negative spot test is obtained. The sample is filtered and then NaOH is added to pH 12.

[7] Samples should be filtered immediately on-site before adding preservative for dissolved metals.

[8] Guidance applies to samples to be analyzed by GC, LC, or GC/MS for specific compounds.

[9] Sample receiving no pH adjustment must be analyzed within seven days of sampling.

[10] The pH adjustment is not required if acrolein will not be measured. Samples for acrolein receiving no pH adjustment must be analyzed within 3 days of sampling.

[11] When the extractable analytes of concern fall within a single chemical category, the specified preservative and maximum holding times should be observed for optimum safeguard of sample integrity. When the analytes of concern fall within two or more chemical categories, the sample may be preserved by cooling to 4°C, reducing residual chlorine with 0.008% sodium thiosulfate, storing in the dark, and adjusting the pH to 6–9; samples preserved in this manner may be held for seven days before extraction and for forty days after extraction. Exceptions to this optional preservation and holding time procedure are noted in footnote 5 (re the requirement for thiosulfate reduction of residual chlorine), and footnotes 12, 13 (re the analysis of benzidine).

[12] If 1,2-diphenylhydrazine is likely to be present, adjust the pH of the sample to 4.0±0.2 to prevent rearrangement to benzidine.

[13] Extracts may be stored up to 7 days before analysis if storage is conducted under an inert (oxidant-free) atmosphere.

[14] For the analysis of diphenylnitrosamine, add 0.008% $Na_2S_2O_3$ and adjust pH to 7–10 with NaOH within 24 hours of sampling.

[15] The pH adjustment may be performed upon receipt at the laboratory and may be omitted if the samples are extracted within 72 hours of collection. For the analysis of aldrin, add 0.008% $Na_2S_2O_3$.

[16] Sufficient ice should be placed with the samples in the shipping container to ensure that ice is still present when the samples arrive at the laboratory. However, even if ice is present when the samples arrive, it is necessary to immediately measure the temperature of the samples and confirm that the 4C temperature maximum has not been exceeded. In the isolated cases where it can be documented that this holding temperature can not be met, the permittee can be given the option of on-site testing or can request a variance. The request for a variance should include supportive data which show that the toxicity of the effluent samples is not reduced because of the increased holding temperature.

[38 FR 28758, Oct. 16, 1973, as amended at 41 FR 52781, Dec. 1, 1976; 49 FR 43251, 43258, 43259, Oct. 26, 1984; 50 FR 691, 692, 695, Jan. 4, 1985; 51 FR 23693, June 30, 1986; 52 FR 33543, Sept. 3, 1987; 55 FR 24534, June 15, 1990; 55 FR 33440, Aug. 15, 1990; 56 FR 50759, Oct. 8, 1991; 57 FR 41833, Sept. 11, 1992; 58 FR 4505, Jan. 31, 1994; 60 FR 17160, Apr. 4, 1995; 60 FR 39588, 39590, Aug. 2, 1995; 60 FR 44672, Aug. 28, 1995; 60 FR 53542, 53543, Oct. 16, 1995; 62 FR 48403, 48404, Sept. 15, 1997; 63 FR 50423, Sept. 21, 1998; 64 FR 4978, Feb. 2, 1999; 64 FR 10392, Mar. 4, 1999; 64 FR 26327, May 14, 1999; 64 FR 30433, 30434, June 8, 1999; 64 FR 73423, Dec. 30, 1999]

EFFECTIVE DATE NOTE: At 66 FR 32776, June 18, 2001, § 136.3 was amended by redesignating paragraph (b)(41) as (b)(42) and by redesignating the second paragraph (b)(40) as new (b)(41) and revising it, effective July 18, 2001. For the convenience of the user, the revised text is set forth as follows:

§ 136.3 Identification of test procedures.

*   *   *   *   *

(b) * * *

(41) USEPA. 2001. Method 1631, Revision C, "Mercury in Water by Oxidation, Purge and Trap, and Cold Vapor Atomic Fluorescence Spectrometry." March 2001. Office of Water, U.S. Environmental Protection Agency (EPA-821-R-01-024). Available from: National Technical Information Service, 5285 Port Royal Road, Springfield, Virginia 22161. Publication No. PB2001-102796. Cost: $25.50. Table IB, Note 43.

*   *   *   *   *

§ 136.4 Application for alternate test procedures.

(a) Any person may apply to the Regional Administrator in the Region where the discharge occurs for approval of an alternative test procedure.

(b) When the discharge for which an alternative test procedure is proposed occurs within a State having a permit program approved pursuant to section